THE STRUCTURE OF GLASS
ELECTRICAL PROPERTIES AND STRUCTURE OF GLASS

STEKLOOBRAZNOE SOSTOYANIE
ELEKTRICHESKIE SVOISTVA I STROENIE STEKLA

СТЕКЛООБРАЗНОЕ СОСТОЯНИЕ
ЭЛЕКТРИЧЕСКИЕ СВОИСТВА И СТРОЕНИЕ СТЕКЛА

THE STRUCTURE OF GLASS

Volume 4

ELECTRICAL PROPERTIES AND STRUCTURE OF GLASS

Edited by

O.V. Mazurin

Authorized translation from the Russian by

E. B. Uvarov, B.Sc., A.R.C.S., D.I.C., A.R.I.C.

Springer Science+Business Media, LLC

1965

The Russian text, originally published by Khimia Press in Leningrad in 1964 as the preparatory material for the IV All-Union Conference on the Glassy State, comprises the reports given at a symposium held in Leningrad in December 1962 on the electrical properties of glasses, organized by the I. V. Grebenshchikov Institute of Silicate Chemistry of the Academy of Sciences of the USSR, the Leningrad Regional Administration of the D. I. Mendeleev All-Union Chemical Society, the S. I. Vavilov State Optical Institute, the A. A. Zhdanov Leningrad State University, and the Lensoviet Technological Institute, Leningrad. The introductory chapter, originally published by Goskhimizdat in Leningrad in 1962 as Chapter II of the Leningrad Technological Institute Trudy Volume LXII, has been revised and updated by the author.

Library of Congress Catalog Card Number 58-44503

ISBN 978-1-4899-5659-0 ISBN 978-1-4899-5657-6 (eBook)
DOI 10.1007/978-1-4899-5657-6

Originally published by Consultants Bureau Enterprises, Inc. in 1965.
Softcover reprint of the hardcover 1st edition 1965

CONTENTS

INTRODUCTION

I. GENERAL ASPECTS OF THE STUDY OF ELECTRICAL PROPERTIES OF GLASSES

II. INVESTIGATION OF GLASSES WITH IONIC CONDUCTION

III. INVESTIGATION OF GLASSES WITH ELECTRONIC CONDUCTION

IV. ELECTRODE PROPERTIES OF GLASSES

INTRODUCTION

PREFACE TO THE AMERICAN EDITION

This is the chapter of the book "Electrical Properties of Glass" concerned with electrical conductivity of solid glasses (with the exception of semiconductor glasses). The Russian edition of the book was published during the first half of 1962.

In preparing this chapter for publication in the United States of America, only minor modifications have been introduced into the main text. However, brief references to new work on the electrical conductivity of glass (much of which is reported in the papers that comprise the main body of this volume) will be found in numerous footnotes.

I hope that this presentation will be of interest to English readers, particularly since an attempt has been made to generalize extensive experimental data which to a very large extent are concentrated in publications little known outside the Soviet Union.

O. V. Mazurin

February 25, 1965
Leningrad

GLASS IN A DIRECT ELECTRIC FIELD*

Nature of Carriers of Electricity in Glass

The first publication concerned with the nature of conduction in glass appeared a little more than 100 years ago [161]. Since then numerous investigators have studied this interesting problem.

Without considering individual studies in detail, let us examine the principal results of these extensive investigations.

The great majority of investigators used high-alkali glasses in their experiments – either pure soda glasses or soda glasses containing small amounts of potassium oxide. If a direct current is passed through such glasses, with the aid of materials containing sodium ions or atoms (sodium amalgam, fused sodium salts, aqueous solutions of sodium salts) as anodes, the passage of current is accompanied by transfer of sodium ions through the glass. The process conforms strictly to Faraday's law.

The most accurate determinations (average deviations from the theoretical values not in excess of 0.5%) were carried out by Warburg [241] and Kirby [189]. Warburg showed that when 6/7 of the Na^+ initially present in the glass had been replaced by electrolysis, all the physicochemical properties of the glass remained unchanged. Most investigations were carried out at 200-300°C, but Quittner [208] investigated several glass compositions at 52°C. Finally, Faraday's law was verified by Tubandt's method [41, 125]. The following procedure was used. A known quantity of electricity is passed through three plates of the material, in intimate contact with thoroughly ground surfaces. The sign and equivalent weight of the electricity carriers are determined from the change in the weight of the plates.

Most of the experiments on verification of Faraday's law were carried out in such a manner that the glass did not undergo any changes during the experiment. In the Tubandt technique the conditions for transfer of charged particles from one plate to another are determined primarily by the state of the facing plate surfaces. The composition of these surfaces apparently remains unchanged during the experiment. Therefore in all these investigations the properties of glass as such were studied, without any disturbing effects due to the electrodes. If an electric current is passed for a long time through glass with the aid of an electrode free ions or atoms of the corresponding alkali metal (mercury, graphite, gold, etc.), a layer with a deficiency of alkali-metal ions begins to form at the anode; the composition and resistance of this layer differ greatly from those of the original glass. The properties of this layer of poor conductivity (often determining the electrical properties of the specimen as a whole) depend not only on the properties of the original glass but to a very great degree also on the properties of the electrode and the conditions of current flow through the specimen. Here we are considering the electrical properties of glass unchanged by electrolysis; i.e., we are concerned with the characteristics of glass which are independent of processes developing in the regions near the electrodes. Effects near the electrodes are highly complex and in most cases have been studied very little. However, they can be disregarded in many practical problems.

The work of Kosman's school [62, 63] cast doubt on the results of experiments on verification of Faraday's law. It was found that Faraday's law is not obeyed when a current is passed for a long time through window glass.

It must be pointed out, however, that the electrodes used in Kosman's work did not contain sodium ions. Therefore the effects observed by Kosman et al. are associated with the formation of a poorly conducting layer

---

*This introductory chapter is taken from O. V. Mazurin, Electrical Properties of Glass, Chapter II, Tr. Leningr. Tekhnol. Inst. im. Lensoveta, Goskhimizdat, Leningrad (1962).

where electronic conduction is quite probable at high field strengths. Faraday's law is always strictly obeyed if the electrodes are chosen correctly. The results obtained by the Kosman school are discussed in greater detail in [73, 107].

Recent experimental results lead to the conclusion that the conduction of alkali glasses from the softening temperature down to the region of room temperatures (at least for glasses whose resistivity does not exceed 10^{13}-10^{14} ohm-cm in this range) is due to alkali-metal ions.*

The next group of studies of the electrolytic conduction of glass is concerned with the possibility of introducing, by electrolysis, ions not present in the original glass. In assessing the results of these investigations we must bear in mind that the conditions for transfer of various metal ions into glass from salt melts, from amalgams, and from salt solutions are very different. This accounts for the apparent contradictions between the results of different investigations. Certain salt melts (in particular, salts of alkali metals) strongly attack glass surfaces. If the rate at which the layer near the anode, deficient in alkali-metal ions, is formed is lower than the rate of glass corrosion, investigations of current versus time relations may lead to the erroneous conclusion that electrolytic substitution of ions proceeds without hindrance. If salt melts are used, there is no doubt that the nature of the alkali-metal salt used for the experiment and the experimental temperature influence the results. If the salt used has a more corrosive effect on the glass, it might be concluded that electrolytic substitution occurs more freely. We therefore find reports in the literature stating that K^+ ions [180, 190] and Li^+ ions [214] pass readily into soda glass, while others state that K^+ ions [157, 214] and Li^+ ions [180, 190] enter soda glass with great difficulty. Evidently the most valuable results are those of experiments in which the electrodes used had little or no corrosive effect on glass (amalgams or aqueous salt solutions). It must be borne in mind, of course, that amalgams contain metals predominantly in the form of atoms rather than ions, so that the conditions for their transfer into glass may be different. However, replacement of a molten salt by an amalgam does not alter the transfer of sodium into soda glass [190].

Unfortunately, most investigators do not give information on the resistances of the circuits used. Nevertheless, in all the publications cited above it is noted that lithium and potassium pass with greater difficulty than sodium from amalgams into soda glass. Tegetmeier's data [231] indicate that the resistance of a circuit with soda glass remains constant in time when a sodium amalgam electrode is used, and increases if the electrode is an Li amalgam. Finally, Quittner's results [208] also show that sodium ions pass more easily than lithium ions from salt solutions into soda glass.

It follows that ions which are either larger or smaller than sodium ions penetrate into soda glass with difficulty; yet potassium ions penetrate readily into potash glass [243].† Lithium ions enter lithium glass from salt melts with relative ease, but after one hour of electrolysis the glass cracks [180]; this confirms yet again that lithium salt melts have a corrosive effect on silicate glasses.

Of the other ions, silver [190] and ammonium [180], and in all probability hydrogen [208], penetrate most easily into glass. Bivalent-metal ions cannot be introduced into glass to any appreciable extent from amalgams [214], although some can be introduced from salt melts [181].

In silica glass and in crystalline quartz [242] current is carried, at least partially, by alkali-metal impurities [226]. This, in particular, accounts for the poor reproducibility of conductivity data for silica glasses (see Fig. 12). The same probably applies to glassy boric anhydride. However, the conduction of alkali-free glasses containing considerable amounts of bivalent-metal oxides cannot be attributed to alkali-metal impurities. For barium borate glasses this was first demonstrated by Markin and Myuller [97]. In all probability this result can be extended to any other alkali-free glass containing over 20-30% of bivalent-metal oxides (at

*Recently Pronkin [265] showed experimentally that the transport number of the sodium ion in alkali aluminosilicate glasses at 80-110°C is $n = 1.00 \pm 0.1$. For discussions of the nature of conduction in alkali aluminosilicate glasses see also [49, 263, 266].

†Comparison of data on self-diffusion of alkali-metal ions with the as yet meager data on heterodiffusion [251] also appears to indicate that alkali-metal ions differing in size from the main alkali-metal ion in the glass have to overcome high energy barriers on entering the glass.

least oxides of heavy metals, such as Ca, Ba, Sr, Pb, and Cd). Even deliberate introduction of considerable amounts of alkali (see Fig. 48) has very little effect on the conductivity of such glasses. Therefore the main constituent of the glass is responsible for transfer of electricity in such cases.

Unfortunately, investigations of the nature of conduction in alkali-free glasses involve very serious difficulties, due primarily to the high resistivities of such glasses. Two recent publications on this subject [32, 34] do not give sufficient cause for regarding the problem as solved (especially since the authors reach directly opposed conclusions). Therefore, in studying the electrical properties of alkali-free glasses we must provisionally accept two alternatives – conduction due to transfer of bivalent-metal ions, and conduction due to movement of electrons. It is possible that ionic conduction predominates in some glasses of this group, and electronic conduction in others. To distinguish it from the well-understood mechanism of alkali conduction in high-alkali glasses, we will use the term "nonalkaline" to describe the conduction of alkali-free glasses.

Effects of Temperature and the Thermal History of the Specimens on the Electrical Conductivity of Glass

Frenkel' [148], Jost [187], and Skanavi [137], on the basis of elementary kinetic concepts, formulated a quantitative theory of ion migration in ionic crystals; this was extended by Stevels [222], Myuller [109], and others to glasses. The derivation presented below is closest to the variant proposed by Skanavi [137].

Each ion at a lattice point undergoes thermal vibrations, the energy of which varies (fluctuates) with time. The probability that the energy of thermal vibrations exceeds the energy U_d binding the ion in the lattice point is equal to $e^{-U_d/kT}$, where k is the Boltzmann constant and T is the absolute temperature. In accordance with the usual chemical terminology, we describe the exit of an ion from a lattice point as dissociation, and U_d as the dissociation energy. The dissociated ion may migrate in the lattice interstices, overcoming the corresponding energy barriers U_a. Ion migration along vacancies, also activated, is not excluded.

In absence of an electric field ion displacement is equally probable in all directions. When a field is applied, the probability of displacement is somewhat greater with than against the field, and a current flows as a result.

Let us consider the quantitative derivation for motion of dissociated ions along interstices (the derivation for migration of vacancies is analogous).

If 1 cm^3 of a crystal contains n_0 ions, the number of ions n having, at each given instant, energy in excess of U_d will be

$$n = n_0 \cdot e^{-U_d/kT} \qquad (1)$$

All these ions will be thrown from lattice sites into the interstices, i.e., they will be dissociated. However, only an ion which is situated in an interstice adjacent to a vacant site will be capable of recombination after decrease of the reserve of kinetic energy. Therefore, for each dissociated ion the probability W of remaining in the dissociated state is equal to the ratio of occupied to vacant sites:

$$W = \frac{n_0 - n_d}{n_d} = \frac{n_0}{n_d} - 1$$

where n_d is the number of vacant sites, equal to the number of dissociated ions, in 1 cm^3. Since $n_0/n_d \gg 1$, we have

$$W = \frac{n_0}{n_d}$$

Thus, the number of dissociated ions n_d is equal to the number of ions having energy in excess of U_d multiplied by the probability of the ion remaining in the dissociated state after its kinetic energy has fallen to the usual level:

$$n_d = nW = n_0 \cdot e^{-U_d/kT} \cdot \frac{n_0}{n_d}; \qquad n_d^2 = n_0^2 \cdot e^{-U_d/kT}; \qquad n_d = n_0 \cdot e^{-U_d/2kT} \tag{2}$$

A somewhat more rigorous derivation of Eq. (2) was proposed by Wagner and Schottky [220, 240], and independently of them by Jost [187].

In absence of a field, the number of displacements per second of dissociated ions in 1 cm^3 of the glass is equal to the number of attempts 2ν of an ion to overcome the energy barrier U_a (ν is the thermal vibration frequency, and an ion raises its kinetic energy to a maximum twice per period) multiplied by the probability of the ion reaching the energy U_a and by the number of dissociated ions n_d. Since all directions are equally probable, the number of displacements n_a in any chosen direction must be one sixth of the total:

$$n_a = \frac{n_d}{6} \cdot 2\nu \cdot e^{-U_a/kT} = \frac{n_d}{3} \cdot \nu \cdot e^{-U_a/kT} \tag{3}$$

When a field is applied, the energy of the ion is somewhat increased when it moves with the field, and somewhat decreased in motion against the field. The accelerating or retarding field strength ΔU is equal to the field strength E multiplied by the ionic charge q and by the distance $\delta/2$ between the equilibrium position of the ion and the potential barrier (δ represents the distance between two potential wells):

$$\Delta U = \frac{qE\delta}{2} \tag{4}$$

Accordingly, $(n_d/3)\,\nu \cdot e^{-(U_a-\Delta U)/kT}$ ions are displaced in the direction of the field and $(n_d/3)\nu \cdot e^{-(U_a+\Delta U)/kT}$ against the field in 1 cm^3 per second. Therefore the excess number of ions Δn transported per second through the potential barrier U_a in the direction of the field must be

$$\Delta n = \frac{n_d}{3}\nu\left[e^{-(U_a-\Delta U)/kT} - e^{-(U_a+\Delta U)/kT}\right] = \frac{n_d}{3}\nu \cdot e^{-U_a/kT}\left(e^{\Delta U/kT} - e^{-\Delta U/kT}\right)$$

For not very strong fields (up to thousands of volts per centimeter) $\Delta U \ll kT$. In this case, expanding the exponential functions in brackets into series, we can confine ourselves to the first degree of approximation:

$$e^{\Delta U/kT} - e^{-\Delta U/kT} = \frac{2\Delta U}{kT}$$

As a result we obtain

$$\Delta n = \frac{n_d}{3} \cdot \frac{2\Delta U}{kT}\nu \cdot e^{-U_a/kT} = \frac{n_d q\delta E\nu}{3kT} \cdot e^{-U_a/kT} \tag{5}$$

We find the specific conductance $\varkappa$ by multiplying Δn by the ionic charge q and the distance δ traveled by an ion in a single displacement, relating the product to unit field strength, and then substituting according to Eq. (2):

$$\varkappa = \frac{(\Delta n)q\delta}{E} = \frac{n_d q^2\delta^2\nu}{3kT} \cdot e^{-U_a/kT} = \frac{n_0 q^2\delta^2\nu}{3kT} \cdot e^{-[(U_d/2)+U_a]/kT} \tag{6}$$

In order to establish the extent to which the above concepts are applicable to glasses, we compare the results derived from them with experimental data.

The exponent of formula (6) for electrical conductivity contains two quantities, U_d and U_a, which cannot at present be either calculated or measured separately with satisfactory accuracy (except for certain simple halide crystals [178]). It is therefore convenient to combine them into a single quantity. This can be done in two ways: (a) $U_0^{(a)} = U_d/2 + U_a$ and (b) $U_0^{(b)} = U_d + 2U_a$. The exponential term in formula (6) is then written as

(a) $e^{U_0^{(a)}/kT}$ or (b) $e^{U_0^{(b)}/2kT}$ If $U_d \gg 2U_a$, variant (b) is preferable, because U_0 in this case is close in magnitude to the dissociation energy. If U_d and $2U_a$ are of the same order of magnitude (which is more probable), the two variants are of equal merit for isolated conductivity investigations.

However, for analysis of the relations between the variations of dielectric losses and conductivity with temperature (see, e.g., [183, 233]) it is important that the figures characterizing the values of U_0 in the expressions for conductivity and dielectric losses should be comparable. Nevertheless, in calculations of U_0 for dielectric losses in the general case only variant (a) is permissible [variant (b) has no convincing basis for most types of losses and is not used in any publication known to us on dielectric losses in glasses].

We shall therefore adhere strictly to calculations in accordance with variant (a), converting as necessary the data of authors using variant (b) for their calculations. For qualitative interpretation of conductivity data there are no additional difficulties in passing from one method of calculating U_0 to the other because the ratios of the energy quantities for glasses of different composition remain the same whatever method is used for calculating U_0, while comparison of data on conductivity and losses is greatly simplified.

Both variants of the expression for U_0 are to be found in the scientific literature on the conductivity of dielectrics. Therefore, in comparing the results of different investigators it must be remembered that $U_0^{(b)} = 2U_0^{(a)}$

We can now rewrite Eq. (6) as follows:

$$\varkappa = \frac{n_0 q^2 \delta^2 \nu}{3kT} e^{-\frac{U_0}{kT}} = A \cdot e^{-\frac{U_0}{kT}} \tag{7}$$

Myuller [109] calls $2U_0$ the energy value Ψ_Φ. For simplicity, we will call U_0 the effective activation energy, remembering that

$$U_0 = \frac{U_d}{2} + U_a \tag{8}$$

Let us now consider the relation between the dissociation and activation energies. According to Myuller [109], the activation energy should be regarded as very small in comparison with the dissociation energy. We do not consider that the arguments in favor of this are sufficiently convincing. According to the calculations of Mott and Littleton [204] for NaCl, the dissociation energy $U_d = 2.9$ eV and the activation energy $U_a = 0.47$ eV. Since Eq. (6) contains the dissociation energy divided by 2, we must conclude that even in the case of the simple NaCl lattice the activation energy can make a substantial contribution to the total effective activation energy $U_0 = U_a + U_d/2$. The role of activation energy is likely to increase for more complex lattices. Indeed, according to Ioffe and Shaposhnikov [46], $U_d/2$ for alkali-metal ions in the quartz lattice is even less than U_a.

The fact that crystallization of simple alkali silicates doubles the effective activation energy [260] also indicates that the activation energy may play a considerable part in the total energy balance. The main effect in crystallization of such silicates is ordering of the lattice. This cannot cause a large increase of dissociation energy because the distance between oxygen and alkali-metal ions does not alter appreciably, while their charges remain constant. On the other hand, the activation energy, which is determined by the ability of an ion to move among other ions in the lattice, can increase greatly with increasing order in the positions of these ions.

Now, if U_d and U_a (and therefore U_0) do not change with temperature Eq. (7) leads to the following relation between specific conductance and temperature:

$$\log \varkappa = a - \frac{b}{T} \tag{9a}$$

or, with specific resistance, which is more convenient from the practical standpoint, we have the expression

$$\log \rho = -\log \varkappa = -a + \frac{b}{T} \tag{9b}$$

Experiments show that for the great majority of glasses the conductivity—temperature relation conforms to Eqs. (9). Since plots of log ρ versus 1/T (or of log $\varkappa$ versus 1/T) are linear, temperature—conductivity relations of glasses are always plotted in these coordinates.

Linearity of plots of log ρ versus 1/T is observed over a region bounded at the high-temperature end by t_g (see [74] for information on the variations of ρ with temperature for glasses over a wide range of temperatures, covering the solid, softened, and molten states) and at the low-temperature end by temperatures at which the specific resistance begins to exceed 10^{14}-10^{16} ohm-cm (see p. 13). Let us examine the relationships characteristic for this linear region, which is of great interest both for theory and for practice.

Since $b = MU_0/kT$,* it is evident that we can calculate the total activation energy U_0 from Eqs. (9), which give the analytical expression for the effect of temperature on the resistance of glass, or from plots of log ρ versus 1/T (in this case from the slope of the line). However, we can do this only if U_0 does not vary with temperature. It may seem at first sight that the fact that Eqs. (9) are obeyed is sufficient guarantee of this. However, it can be easily shown [146] that if U_0 varies linearly with temperature Eqs. (9) remain valid. Indeed, let $U_0 = U_C \pm U_T T$. Then

$$\varkappa = \frac{n_0 q^2 \delta^3 \nu}{3kT} e^{-\frac{U_0}{kT}} = Ae^{-\frac{U_0}{kT}} = Ae^{-\frac{U_C \pm U_T \cdot T}{kT}} = Ae^{\pm\frac{U_T}{k}} e^{-\frac{U_C}{kT}} = A_1 e^{-\frac{U_C}{kT}} \tag{10}$$

Evidently, in this case we cannot find the total U_0, but only a part, U_C, from the temperature relation. However, the preexponential term changes considerably. Therefore, the following procedure can be used to determine whether U_0 remains constant with change of temperature.

In contrast to U_0, the preexponential term $A = n_0 q^2 \delta^2 \nu / 3kT$ in Eq. (7) can be approximately calculated. The values of k and q are known to a high degree of accuracy, and δ can be estimated tentatively.

The sum of the radii of Na^+ and O^{2-} ions is 2.34 Å. Assuming that Na^+ ions in the glass are surrounded by coordinated oxygen ions, we can estimate the distance between the centers of two neighboring Na^+ ions as twice the sum of the radii of Na^+ and O^{2-} ions, i.e., 4.68 Å. We take 5Å as a round figure. It must be remembered that δ may vary with the composition, but at present this variation cannot be taken into account. We easily find n_0 assuming that all the alkali-metal ions in the glass take an equal part in the transport of electricity. Although the glass contains a certain proportion of ions not involved in transport of electricity, the error due to this assumption is relatively small and decreases with increasing R_2O content. The temperature taken should be the average value for the region over which the relation was determined.

With regard to ν, it is known that the vibration frequency of ions in the crystal lattice is 10^{12}-10^{13} cps. We take $3 \cdot 10^{12}$ as the average value, taking into account that this frequency may vary considerably in either direction with change of the ion bonding strength.

In view of the approximate character of the values taken for δ and ν, and also of the fact that both of these values can change with the composition (the variations of ν may, of course, be greater than variations of δ, but since the equation contains the second power of δ, variations of δ can also have a considerable influence on A), we can assume that A is determined to a precision of one order of magnitude. It is easy to calculate A if we put q in coulombs (amp-sec), n_0 in cm^{-3} (alkali-metal ions per cm^3), and k in joules/degree (V · amp · sec/deg) into the above expression. We take T = 500°C. Then for a silicate glass containing 20% R_2O we have A = 90 $ohm^{-1} \cdot cm^{-1}$ or log A = 1.95. It is easy to show that introduction of bivalent oxides into an alkali glass does not alter the theoretical values of A which decreases or increases in proportion to the R_2O content. In the transition from silicate to borate glasses A decreases on the average by 1/5 (for equimolecular contents of alkali-metal oxides).

---

*M is 0.4343, the modulus for converting natural to common logarithms.

TABLE 1. Values of log A and U_0 for Certain Alkali Glasses*

No.	Glass composition, mole %	log A_{exptl}	log A_{calc}	U_0 exptl, eV
1	K_2O—15; SiO_2—85	1.55	1.77	0.83
2	Na_2O—40; SiO_2—60	1.57	2.25	0.51
3	K_2O—40; SiO_2—60	2.00	2.25	0.63
4	Na_2O—20; K_2O—20; SiO_2—60	4.18	2.25	1.03
5	K_2O—20: Li_2O—20; SiO_2—60	2.84	2.25	1.05
6	Na_2O—10; PbO—20; SiO_2—70	1.46	1.65	1.04
7	Na_2O—30; PbO—20; SiO_2—50	2.59	2.13	0.82
8	Na_2O—30; BaO—20; SiO_2—50	2.33	2.13	0.86
9	Na_2O—20; BaO—40; SiO_2—40	1.91	1.95	1.10
10	Na_2O—10; B_2O_3—90	1.80	1.55	1.36
11	Na_2O—30; B_2O_3—70	1.52	2.05	0.75
12	Na_2O—10; CaO—10; B_2O_3—80	1.20	1.55	1.17
13	Na_2O—10; MgO—30; B_2O_3—60	1.75	1.55	1.13

*From the data of Mazurin et al.

Table 1 gives values of log A for a number of alkali glasses. It is seen that the experimental values of log A for the most diverse alkali glasses for which the number of carriers of electricity can be determined differ from the calculated values by less than an order of magnitude, with only a single exception (glass 4).* Therefore, leaving aside for the moment the cause of the deviation for glass 4 (see p. 42), we can claim that for the great majority of alkali glasses the experimental and calculated values of the preexponential term are in good agreement. This gives reason to believe that U_0 is almost or entirely independent of temperature,† and we can calculate the total activation energy from the slope of the linear plot of log ρ versus $1/T$. Table 1 gives some values of U_0. It is seen that the slope for different alkali glasses varies by a factor of nearly 3.

Because the values of log A are similar while U_0 varies considerably, when the conductivity–temperature relations of various glasses are plotted in log ρ versus $1/T$ coordinates a system of straight lines converging at high and diverging at low temperatures is obtained (Fig. 1). Although glasses with high log A (for example, glass 4) deviate somewhat from this, the following approximate rules can be formulated for not excessively high temperatures:

1. the higher the U_0 of a glass, the greater is its resistance at the given temperature;
2. the lower the experimental temperature, the greater is the influence of composition on resistance (see, for example, [38]).

An important practical conclusion can be drawn from the fact that log ρ is a linear function of $1/T$. The temperature–resistivity relation in the temperature range in question can be regarded as completely defined if we know either the resistance at two different temperatures or the values of the constants a and b in Eqs. (9) (or, what amounts to the same thing, the values of log A and U_0).

So far we have been discussing alkali glasses. Table 2 gives experimental values of log A for certain alkali-free glasses. For these glasses the values of log A show large differences from one composition to another. At the same time, the values of log A are much lower. On the assumption that current is carried in alkali-free glasses by bivalent-metal ions, the average values of A correspond to amounts of current carriers,

*If, as proposed in [259], γ is taken into account, log A_{calc} is decreased by 0.7 of an order of magnitude, which, on the average, will increase the discrepancy between the calculated and experimental values of log A. However, even in this case the difference will not exceed an order of magnitude for the great majority of glasses.

†As already mentioned, the difference between the experimental and calculated values of A (due to the approximate nature of the calculation) apparently does not exceed an order of magnitude. It is easy to show that at $U_T = \pm 2 \cdot 10^{-3}$ eV/deg A must change by exactly one order of magnitude. This is the maximum possible value of the temperature coefficient of the activation energy. Even in this extreme case the proportion $U_T T$ of the activation energy not taken into account is not very large (at 20°C, $U_T T = 2.06$ eV, and at 230 and 430°C, it is 0.1 and 0.14 eV, respectively).

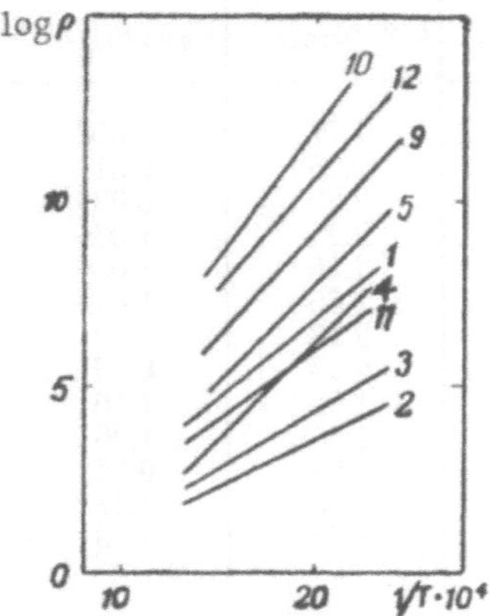

Fig. 1. Temperature–resistance relations for certain glasses (from the data in Table 1 for the corresponding numbers).

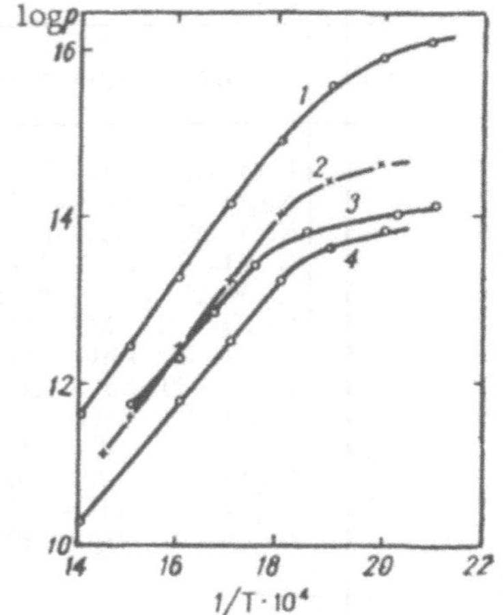

Fig. 2. Temperature–resistance relations for certain alkali-free glasses, from the data of Odelevskii and Verebeichik [124]. 1) Barium aluminosilicate glass; 2, 3, 4) glasses in the system $CaO-MgO-Al_2O_3-SiO_2$.

which constitute approximately $^1/_{100}$ of the contents of bivalent-metal ions in the glasses (it should be taken into account that if ν changes in the transition from alkali-metal to nonalkali ions, the change can be only an increase).

It is stated in some publications on the conductivity of glasses [179, 222] that the expression

$$\log \varkappa = A + BT + CT^2 \tag{11}$$

where T is the absolute temperature, and A, B, and C are constants, represents the effect of temperature more accurately than Eqs. (9) for certain glasses.

Although we have no cause to doubt the accuracy of the measurements reported by the authors cited above, we must point out that the conductivities of glasses of similar composition, according to their own data, cor-

TABLE 2. Experimental Values of log A and U_0 for Certain Alkali-Free Glasses*

Glass composition, mole %	log A	U_0, eV
SiO_2—50; PbO—50	+0.6	1.08
SiO_2—60; BaO—40	—0.3	1.25
SiO_2—50; BaO—50	0.0	1.24
SiO_2—50; CaO—50	+1.1	1.44
SiO_2—50; PbO—40; CaO—10	+0.8	1.20
SiO_2—50; PbO—30; CaO—20	+0.5	1.29
SiO_2—10; PbO—50; Al_2O_3—40	+0.4	1.37
SiO_2—55; CaO—30; Al_2O_3—15	+0.9	1.73
SiO_2—77.3; B_2O_3—9.3: As_2O_3—0.7; Al_2O_3—12.7	+0.3	0.92
B_2O_3—60; CaO—40	—1.2	1.44
B_2O_3—60; PbO—40	+1.7	1.41
B_2O_3—60; BaO—40	+1.4	1.64

*From the data of Mazurin et al.

respond to Eq. (9a). Our attempts to reproduce their results proved unsuccessful. A temperature relationship represented by Eq. (9a) was found for all the glasses. *

Most investigations of the temperature—resistivity relation have been confined to the region up to 10^{11}-10^{12} ohm-cm. Few determinations [2, 33, 49, 96, 124, 155, 179a, 222, 235] have been reported at higher resistivities. According to most of the results, at resistivities above 10^{14}-10^{15} ohm-cm the temperature coefficient of conductivity decreases (Fig. 2). However, in some cases [33, 155, 179a] the relation remained linear up to 10^{16}-10^{17} ohm-cm. We obtained a similar result for certain alkali-free glasses in a joint investigation with Zorin [253].

The inflections on the plots may be due to two causes. One may lie in the use of an incorrect method of measurement. The rate of polarization processes in glass depends strongly on the temperature [137, 274, 233]. The rate is high at high temperatures; no matter how soon after the potential has been applied in the conductivity measurement, polarization has already been completed. At low temperature polarization is established slowly (see p. 16) and the current due to it is added to the current of through conduction which we are trying to measure. As a result, the values found for the current are too high and hence low values are obtained for resistivity [253].

A second cause of the breaks in the plots may be associated with changes in the nature of the conduction (from conduction due to the principal lattice ions to impurity conduction) and in the nature of the electricity carriers (for example, transition from ionic to electronic conduction). Nearly all halide crystals give linear plots with breaks; most investigators consider that these breaks are due to transition to impurity conduction [137, 178].† Nevertheless, the concept of an "impurity" ion is itself not quite clear in the case of glass. In the absence of a regular lattice it seems impossible to distinguish between principal and impurity ions. The suggestion that at low temperatures ionic conduction is replaced by electronic [49] appears to be quite probable. However, nothing definite can be said about the nature of these breaks until careful investigations have been carried out.

Let us examine briefly the influence of the thermal history of the specimen on conductivity (a detailed discussion of this subject is contained in [74]). Nearly all the literature data available at this time refer to the effects of annealing and chilling on the conductivity of alkali glasses. We shall therefore confine our discussion to glasses with alkali conduction.

It is well known that chilled glasses have higher conductivities than annealed [43, 194, 215]. A detailed bibliography is given in [74]. The changes taking place during repeated chilling and annealing are reversible and therefore cannot be associated with crystallization processes. On the whole, the variations of conductivity as the result of heat treatment are quite analogous to the corresponding variations of other physicochemical properties. With the usual types of annealing and chilling, the difference between the logarithms of the resistivities of chilled and annealed specimens is in the range of 0.3-0.5. With the aid of special techniques the difference can be increased to an order of magnitude and over.

There are various views concerning the nature of this effect. We accept the views put forward in most detail in the USSR by Kobeko [54]. They can be summarized as follows.

It may be assumed that in any liquid (this has been demonstrated by x-ray diffraction for water) the relative arrangement of as well as the distances between the individual particles alters with temperature. Despite all the possible deviations of a fluctuational character, the statistical mean distribution of the particles is strictly constant at any given temperature, and is characteristic for that temperature only. In the case of silica melts it is probable that at the highest temperatures the amorphous network is closest in structure to cristobalite, and a gradual transition to a tridymite-like and finally to a quartz-like structure takes place with decrease of temperature. The statistical mean order of the ions may also become greater with fall of temperature because of the weaker influence of thermal motion. In one way or another, the melt structure must become more compact and more ordered with decrease of temperature.

---

*According to Lengyel et al. [271], magnesium barium silicate glasses containing two alkali ions deviate from this relationship at temperatures above 350-400°C.

†A different point of view is put forward by Myuller [114].

The structure (relative positions of the particles in the network) of solid glass should not change with temperature. When the temperature of solid glass is changed, the result is merely a certain change in the distances between the particles while their relative positions remain strictly constant. Consequently, solid glass retains the structure characteristic of the softening range. It is evident that chilled glass should retain the structure corresponding to higher temperatures, while the structure retained by annealed glass should correspond to lower temperatures. Since the low-temperature structure should be more compact and more ordered, annealed glasses should have lower conductivity (and also higher density, higher refractive index, etc.). It is important that it is possible on this basis to interpret from a unified standpoint the effects of annealing and chilling on the most diverse physicochemical properties of glass [74].

Chilling has little or no effect on the conductivity of the majority of alkali-free glasses [124, 206a]. We have also found that chilling has only a very slight effect on the conductivity of alkali-free glass used for production of glass fibers.

Odelevskii and Verebeichik [124, 126] noted that chilling has an unusually strong influence on the electrical properties of certain alkali-free glasses.*

Effect of the Time of Action of Direct Voltage on the Conductivity of Glass†

If a voltage is applied to a dielectric, the charge induced owing to the capacity of the dielectric grounded rapidly, and the current then measured, it is found that in many cases the current decreases with time. This decrease of current with time was investigated in the now classical studies of the Ioffe school [46]. The most important conclusions of these studies are summarized below.

1. At the instant the voltage is applied, the distribution of potential over the thickness of the dielectric is uniform (because macroheterogeneous specimens are taken for the experiments). However, only a fraction of a second after application of the voltage the field in the dielectric becomes distinctly nonuniform. Figure 3 shows two different types of potential distribution curves observed in the investigations of the Ioffe school. It was shown that this kind of potential distribution is caused by accumulation of volume charges at the electrodes, apparently as the result of changes in the concentration of the charge carriers of the appropriate sign in one or both (dependent on the nature of the charge carriers) electrode regions.

The field in the dielectric is curved even when the volume charge density is constant over the volume. The field is distorted still further with change of charge density. The regions of existence and the signs of the volume charges are indicated in Fig. 3. It must be pointed out that in all the crystals studied, with the exception of quartz, the thickness of the layer in which the volume charge is concentrated is very small, on the order of a few microns.

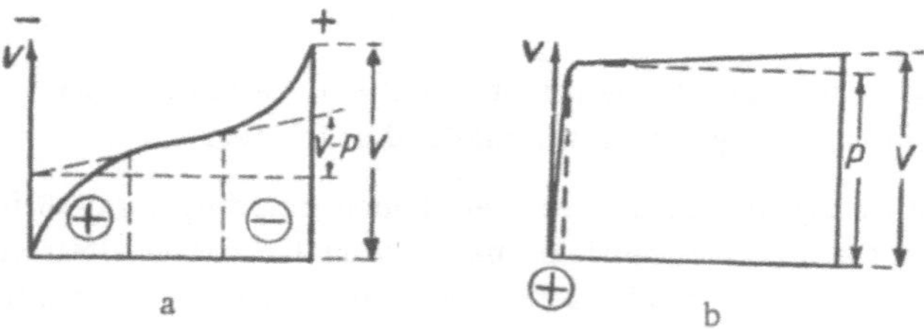

Fig. 3. Potential distribution in crystals. a) Quartz; b) calcite. Regions of volume charge formation are indicated by vertical dash lines. The signs of the volume charges formed are shown in the circles [30].

*However, investigations carried out by Z. A. Levtsova (see [250]) and Zorin [253] jointly with Mazurin suggest that the effects detected by Odelevskii and Verebeichik are associated primarily with surface conductivity, which may play an important part in glasses of the compositions investigated at temperatures far above 100°C.

†A somewhat different viewpoint on the problems discussed here is presented in [246]. Some aspects are discussed in [254].

2. The formation of a volume charge gives rise to a back emf P. The volume charge, and therefore the value of P, increases as the current passes through the dielectric. Assuming that the resistance R of the crystal does not alter during the experiment, we can represent the current I as follows:

$$I = \frac{V - P}{R} \tag{12}$$

where V is the voltage initially applied.

Thus, the current decreases continuously throughout the time of accumulation of the volume charge. When the volume charge has reached a certain limit, dependent on the type of crystal and the magnitude of the applied voltage, P reaches a maximum (P_{max}). Residual current, unchanging with time, passes through the specimen:

$$I_{res} = \frac{V - P_{max}}{R}$$

When the external voltage is removed ($V = 0$), only the emf P acts in the specimen. If the electrodes are short-circuited, current passes through the specimen in the opposite direction under its influence. The flow of current is accompanied by dissipation of the volume charge and therefore by gradual decrease of P and I. If the formation and dissipation of the volume charge follow the same laws, the curves representing variations of current flow in the two directions with time must be the same; this is found to be the case. P may reach high values, and therefore the term "high-voltage polarization" is often used to describe the decline of current in a solid dielectric.

3. Evidently the value of the current I cannot be a measure of the conductivity of the dielectric itself. The total current passing through the dielectric is determined more by the processes taking place at the regions of contact between the dielectric and the electrodes than by the properties of the dielectric itself. It is obvious that if materials capable of compensating partially or completely the loss of conducting particles at the electrodes were used for the electrodes, I would change substantially.

The problem was to determine, with the use of any electrodes, the true resistance R_{true}, the resistance characteristic of a specimen not yet changed by the processes occurring in it under the influence of an electric current, i.e., a specimen without volume charge. Ioffe proposed several methods for this purpose. They form two main groups: a) determination of the current I_0 immediately after application of the voltage, before the volume charge has had time to form; b) determination of P immediately after removal of the voltage, before the volume charge has had time to decrease considerably. In either case we should find R_{true}:

$$R_{true} = \frac{V}{I_0}, \quad R_{true} = \frac{V - P}{I} \tag{13}$$

We must point out that Eqs. (13) must give correct results only if during the time needed for the first readings of I and P the steady state has not been disturbed, i.e., if the volume charge has not started to form (case a) or to dissipate (case b). This condition, if satisfied, would correspond to the shape of the time–current plot shown in Fig. 4. In this case the formation of volume charge could be disregarded up to time t'. However, the shape of this curve is not characteristic for glass, and Eq. (13) can therefore not be used.

Apparently the most reliable and simplest method for detecting R_{true} of dielectric crystals would be by measuring the decrease of potential in a crystal through which a current is flowing with the aid of probes placed at a sufficient distance from the electrodes (outside the sphere of action of the volume charge). This method is widely used in studies of conductivity of semiconductors [47], and could probably be used with equal success for investigating the true conductivity of dielectrics.

We have examined studies of the conductivity of crystals in such detail because, in our opinion, certain investigators studying the electrical properties of glasses apply without any corrections the results obtained by the Ioffe school, based on crystal studies. Nevertheless, the passage of current through glass differs essentially in certain respects from the passage of current through the crystals studied by the Ioffe school.

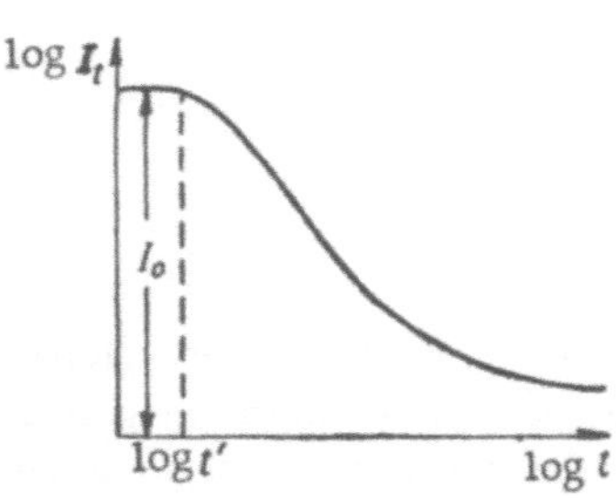

Fig. 4. Rigorous method for determination of the initial current for calculation of R_{true}.

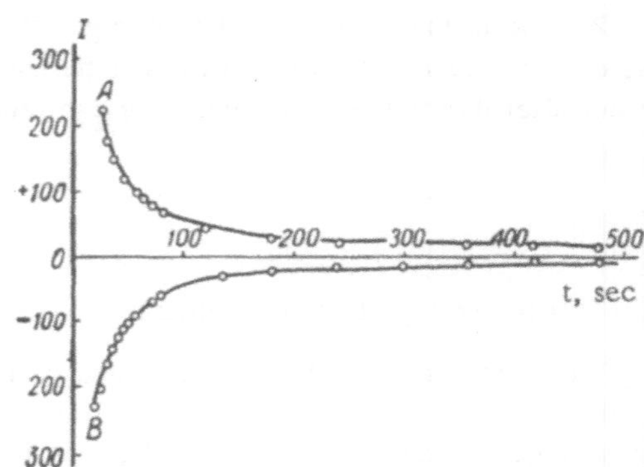

Fig. 5. Typical charge (A) and discharge (B) curves of a glass [170].

Let us now consider the most important experimental data relating to the passage of direct current through glass.*

The nature of the time dependence of current passing through glass depends very much on the temperature. For solid glasses, the entire region of attainable temperatures can be divided into three parts. At low temperatures (at which the resistivity of glasses exceeds 10^{11}-10^{12} ohm-cm) the current gradually decreases with time (Fig. 5); this is analogous to the decrease of current passing through other dielectrics. At higher temperatures the current remains virtually constant from the instant of application of the voltage (more correctly, 10^{-2} to 10^{-3} sec after the application – the time determined by the capabilities of modern apparatus). At even higher temperatures (when the resistivity of glasses is below 10^6-10^7 ohm-cm) the current again decreases with time.

Investigations of potential distribution over the thickness of a specimen are very important. Fairly numerous investigations have shown that at low and moderate temperatures the distribution of potential across the specimen remains linear in experiments lasting one hour and longer [23, 31, and others].† In the high-temperature region the decrease of current is accompanied by a change of the distribution of potential across the specimen. The decrease of current and therefore distortion of the potential curve can be prevented by the use of nonpolarizing (amalgam or salt) electrodes in the high-temperature region [82].

It may be concluded from this that at high temperatures the decrease of current with time (described by some physicists as forming of the specimen) must be caused by processes near the electrode, associated with removal of charge carriers from the layer near the anode. The effects observed here are completely analogous to those investigated by Ioffe. The decrease of current in the low-temperature region is not due to the influence of the electrodes; this is shown by the above-mentioned fact that the distribution of potential in the glass remains linear while the current falls, and by the fact demonstrated by Mazurin and G. P. Nikolina that in the low-temperature region the decrease of current with time is of the same character with both polarizing and nonpolarizing electrodes (Fig. 6).

The general laws of this decrease of current are well known. If a specimen is kept under constant voltage until the current ceases to decrease, the electrodes are short-circuited, and the back current is investigated, it is found that the curves representing the forward and back currents are quite similar. The current–time relation for both the forward and the reverse current passing through glass is best represented by the expression

$$I_t = At^{-n} \tag{14}$$

where I_t is the current at time t, and A and n are constants, with n for glasses usually in the range 0.65-0.8.

It is important that for glass this relation holds down to very short times (down to $2 \cdot 10^{-6}$ sec according to Tank [232]) without any tendency for the dependence of current on time to decrease.

*The first systematic studies of these relationships were carried out by Kurtts [61].

†A similar result was obtained by Zorin and Mazurin.

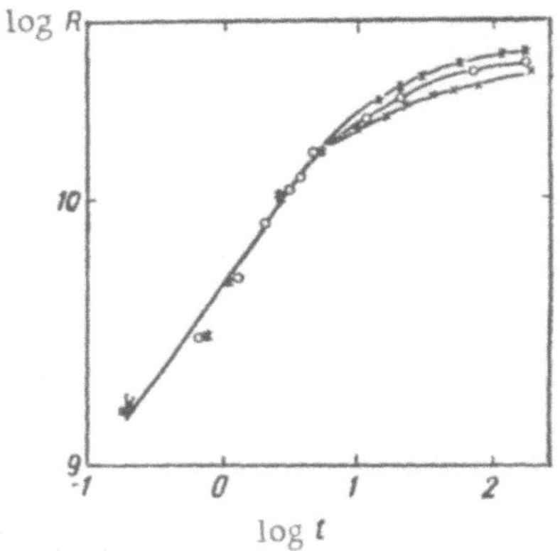

Fig. 6. Resistance versus time curves at constant voltage for glass with the use of different anodes (t in minutes). Chemical laboratory glass No. 29 (10 wt. % Na_2O, 3 wt. % K_2O); specimen in the form of a thin-walled flask. Temperature 43°C. Anode materials: ×) NaCl solution; •) KCl solution; ○) metallic mercury. Data of Mazurin and G. P. Nikolina.

Investigations of the current decay function in dielectrics in the region of short times have been few because of the great experimental difficulties encountered in such determinations. Therefore, in our opinion, the following approach to the effect in question may prove very fruitful.

In the general theory of dielectric losses (see, e.g., [137]) it is shown that the dependence of energy losses on frequency in an alternating field can be calculated from the known function of current decay with time for all cases where the superposition principle is valid.

It is known that the superposition principle holds for glasses in weak electric fields. Therefore, from the solution of the direct problem – calculation of the function $\varepsilon'' = f(\omega)$ from the function $\varkappa = f(\tau)$ (where ω is the angular frequency of the alternating field and ε'' is the dielectric loss factor), carried out by Schweidler [219] – we can solve the reverse problem, that of calculating the function $\varkappa = f(\tau)$ from the function $\varepsilon'' = f(\omega)$ [261].

Schweidler showed that if the current decay function is represented by the formula $i = \beta C_0 \tau^{-n}$, then the active component of the alternating current flowing through the dielectric is expressed as

$$i_a = \omega E_0 \frac{\beta C_0}{\omega^{1-n} \cdot 2\Gamma(n) \cdot \cos[(1-n)\pi/2]}$$

Here C_0 is the so-called zero capacity, determined by the geometrical dimensions of the specimen, β is a constant (for a given dielectric at a given temperature), Γ is the symbol for the gamma function, and E_0 is the peak value of the alternating field strength. Simple but cumbersome transformations (omitted here) lead to the following result.

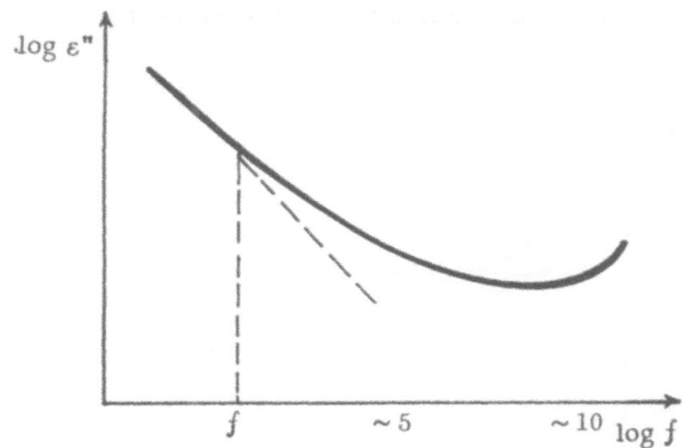

Fig. 7. General character of the dependence of the loss factor of oxygen-containing glasses on frequency. Below the frequency f' dielectric losses are determined almost entirely by conduction losses, which are represented by the dashed line.

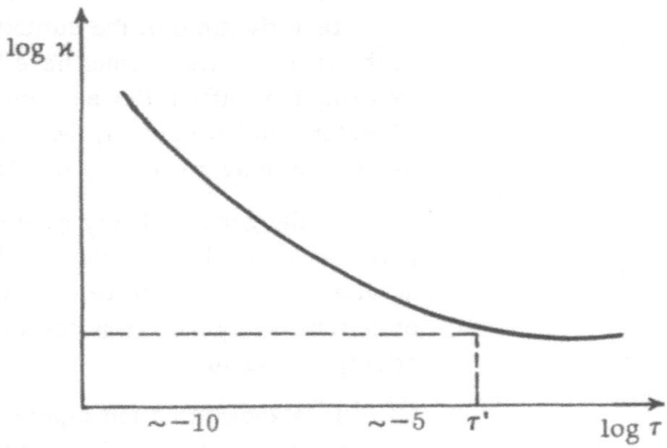

Fig. 8. General character of the dependence of conductivity of oxygen-containing glasses on time from the instant of application of a direct field. After the time τ' ($=1/2\pi f'$) the conductivity remains constant with time.

If the function $\log \varepsilon'' = f(\log \omega)$ is linear, the function $\log \varkappa = f(\log \tau)$ is also linear, and

$$\log \varkappa = \log \left[\frac{\varepsilon'' \omega^{1-n} \Gamma(n) \cdot \cos[(1-n)\pi/2]}{1.8 \cdot 10^{12} \pi^2}\right] - n \log \tau \tag{15}$$

Here $n = 1 + \Delta \log \varepsilon'' / \Delta \log \omega$, where $\Delta \log \varepsilon''$ and $\Delta \log \omega$ are the corresponding changes of both quantities over any arbitrary region of the linear plot.

The curvilinear plot of $\log \varepsilon''$ versus $\log \omega$ can be reduced with any desired degree of accuracy to a sum of straight lines (provided that the curvature of the curve is either positive or zero for all regions). For each of these straight lines the corresponding linear plot of $\log \varkappa$ versus log i is found and these plots are then summed. The resultant curve represents the required relationship.

It has now been established from extensive experimental data that in the frequency range of $1\text{-}10^{10}$ cps the loss versus frequency plot is of the form shown in Fig. 7 (see, e.g., [223, 262]). It is easy to show by means of the relation (15) that the dependence of conductivity on time, the character of which is clear from Fig. 8, corresponds to the curve of Fig. 7. Thus, in glasses of the most diverse compositions, starting not later than

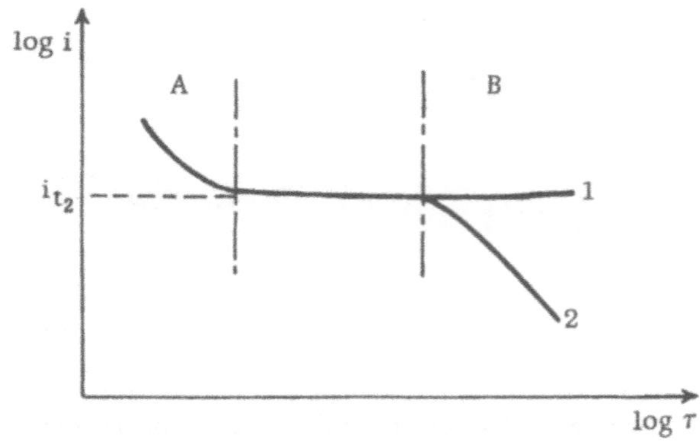

Fig. 9. General character of current variation in glasses with time. A) Region of volume polarization; B) region of polarization at the electrodes. 1) Nonpolarizing electrodes; 2) polarizing electrodes.

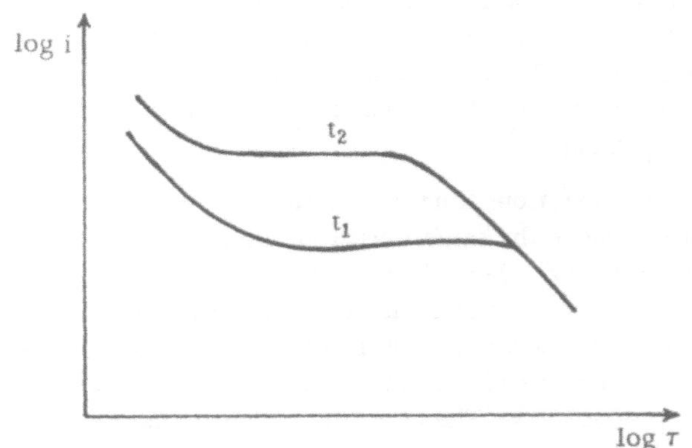

Fig. 10. Variations of the dependence of current on time with change of the experimental temperature.

10^{-11} sec from the instant of application of a direct field, the current flowing through the specimen decreases continuously until all polarization processes are complete in the glass.

As a rule, direct experimental determination of the current can be carried out 10^{-1}-10^{-3} sec after application of the field (see, e.g., [49]). It is evident that even if the influence of transitional processes [253] can be excluded in the determinations, conductivity of glass extrapolated to zero time is an arbitrary value dependent on the speed of response of the apparatus.

The term "true" conductivity (in Ioffe's terminology) should be applied to the conductivity value characterizing motion of charge carriers in the glass in a direct field, independent both of local displacements of charged particles in the glass during establishment of polarization and of processes at the electrodes.

It follows from evaluation of literature data and from a considerable number of experiments carried out by Mazurin, Brailovskii, and Zorin (see, in particular, [245, 247, 261] that the variation of current in a dielectric with time in the general case is of the character evident from Fig. 9. The "true" conductivity of the dielectric can be calculated from the current i_{t_2}.

In the general case, possible superposition of regions A and B in dielectrics is not excluded. If such is the case determinations of the "true" conductivity of the material become impossible without the use of special nonpolarizing electrodes (capable of making up the deficiency of charge carriers in the space adjoining the electrodes). In the case of glasses (at least the great majority of them) the time interval during which the conductivity does not alter, no matter what electrodes are used, is fairly large. It is therefore possible to use graphite, mercury, silver, and other polarizing electrodes for the measurements [82].

The boundaries of regions A and B converge with change of temperature (Fig. 10).

It is evident from the foregoing that the mechanism of polarization processes causing increased current in glasses during the first period after application of the field is fully analogous to the mechanism of dielectric losses in glasses (for a discussion of the latter see, e.g., [223, 274]).

Effect of Composition on the Conductivity of Glass

In examining experimental data on the influence of composition on the conductivity of glass, we should remember that different investigators have obtained different results for the same glasses. This is especially clear from a comparison of data on the conductivity of glass with the composition $Na_2O \cdot 2SiO_2$ (Fig. 11). Most of the results are included in a relatively narrow range of resistivities, but some deviate quite considerably. It must be borne in mind that conductivity data for glasses of this composition show very good reproducibility for different meltings, much better than for glasses of more complex composition.

In some cases these deviations must undoubtedly be attributed to errors in preparation of the glasses or in the determinations (especially in determination of the true temperature of the specimen). The possibility is not excluded, however, that the glass-making conditions, and in particular the temperature of the final melting stage and the rate of cooling from the melting to the working temperature, often play a significant part. The effect of the melting conditions on conductivity of glass has not yet been established, but the possibility of such an effect cannot be excluded (this refers to variations in the structure of glass in accordance with the conditions of preparation rather than to changes in its composition during the melting).

Even with the most careful determinations, discrepancies of up to 100% between the results of different workers must be regarded as common. This is clearly illustrated by comparison of the data of Markin and

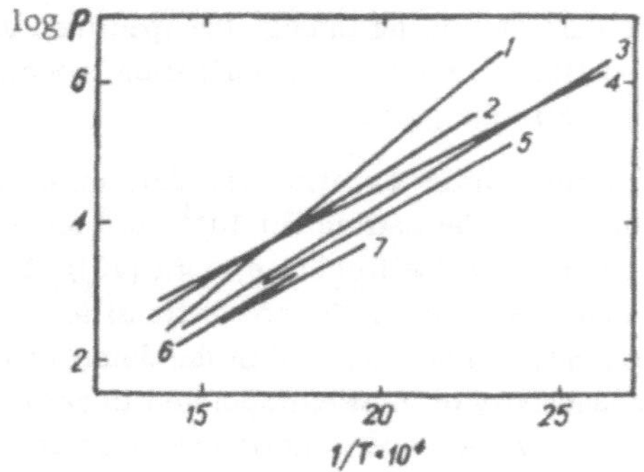

Fig. 11. Resistivity of glass corresponding to sodium bisilicate in composition, from the data of different authors: 1) [168]; 2) [195]; 3) [215]; 4) [87]; 5) [73]; 6) [8]; 7) [216].

Myuller [96] and of Takking and Shchegoleva [143] (see Fig. 14) on the conductivity of glasses in the system $Li_2O-B_2O_3$ (the same carefully thought-out technique was used in both the investigations). Often the deviations can be considerably greater.

The deviations from one melting to another in the investigations of the same worker are usually smaller, in the range 15-35%, dependent on the composition of the glass. This high percentage error should cause no surprise. In fact, in the case of glasses containing 10-15% R_2O, variations of 1% in the contents of certain bivalent oxides alter the resistivity by 50-60%. Therefore the combined use of data reported by different authors is generally permissible only after the results have been checked against data for the same or similar glasses.

It is known that the influence of composition on electrical properties of glasses cannot in general be represented by additive relations. Therefore evaluation of the extensive experimental material now available presents considerable difficulties. It is first necessary to solve the problem of rational principles of glass classification in relation to electrical characteristics. We consider that glasses should be classified in accordance with the nature of the electricity carriers. Nearly all the glasses known at this time can be divided into four groups on this basis.

The first group comprises glasses with alkali conduction (silicate, borate, phosphate, and aluminate high-alkali glasses, and also corresponding low-alkali glasses with little or no bivalent oxides); the second, glasses with nonalkaline conduction (predominantly alkali-free silicate and other oxide glasses, the conduction of which does not depend on transport of alkali-metal ions but which at the same time do not have distinct semiconductor properties). The third group consists of semiconductor glasses (chalcogenide glasses, glasses containing iron or vanadium, and other glasses with considerable electronic conduction). Fluoride glasses, for which anionic conduction may be postulated, now have to be put into a separate group. As already noted, the semiconductor glasses are not considered here.

It is evident that this grouping is somewhat arbitrary, but we consider it to have a number of advantages for a systematic examination of the available material.

All the glass compositions are given in molecular percentages.

Glasses with Alkali Conduction

Silica Glass. The data of different workers on silica glass are plotted in Fig. 12. The great variance of the results is probably due to the fact that current is carried in silica glass by alkali-metal impurities, the contents and bonding of which in this glass depend to a very great extent on the quality of the raw material and the manufacturing process.

Systems R_2O-SiO_2.* Such systems have been studied by various workers. A review of the work of different investigators is given in [76]. As already noted, the results of different workers for the same system usually differ; we therefore give in Fig. 13 the results of our own investigations, in which the conductivities of glasses of three different systems and one cesium silicate glass were determined by the same technique. The following main conclusions can be drawn from the graph. The resistance falls with increasing alkali oxide content (which is quite natural, as current is carried in these glasses by alkali-metal ions). The first additions of alkali oxides appear to have the most effect (according to most investigators, as Fig. 12 shows, log ρ for silica glass at 150°C is 16-17). At moderate alkali contents potash glasses have the highest resistance and lithium glasses the lowest. Variations of alkali oxide content have the greatest effect on potash glasses and the least on lithium glasses.

*For data on the system R_2O-GeO_2 see [255].

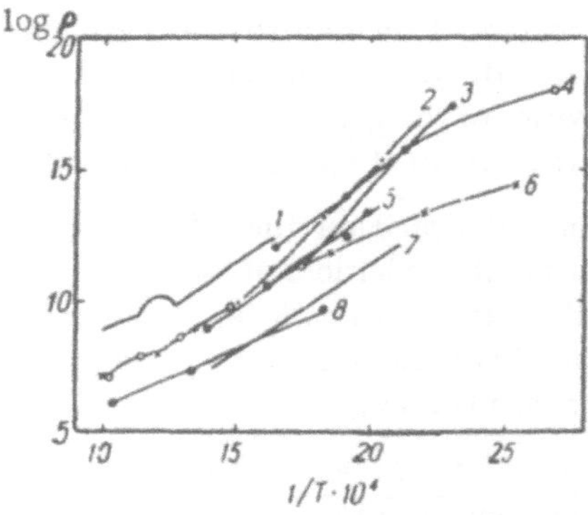

Fig. 12. Resistivity of silica glass, from the data of various authors: 1) Konovalova and Pryanishnikov [129, 131]; 2) Bogoroditskii and Fridberg [11]; 3) data of the Leningrad Physicotechnical Institute [129, 131]; 4) Sosman [218]; 5) Williams [239]; 6) Mikhailov [129, 131]; 7) Strauss [226]; 8) Siemen [217].

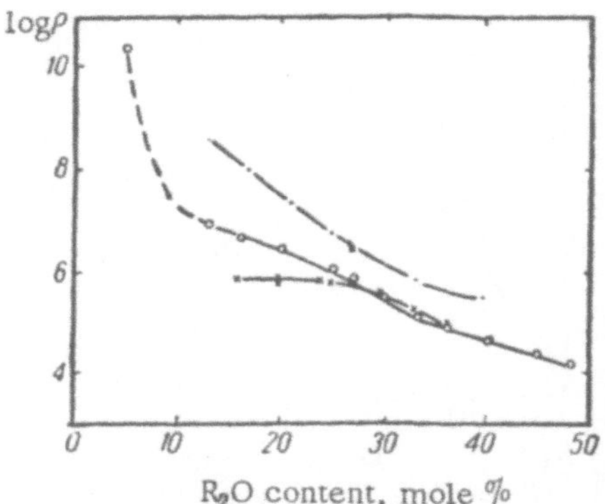

Fig. 13. Resistivity of two-component alkali silicate glasses at 150°C. ×) Li_2O-SiO_2 (Yu. A. Shmidt, Mazurin, and N. G. Suikovskaya, and [76]); ○) Na_2O-SiO_2 ([76] and Mazurin, N. G. Suikovskaya, and K. K. Evstrop'ev); ●) K_2O-SiO_2 ([76]); •) Cs_2O-SiO_2 ([130]).

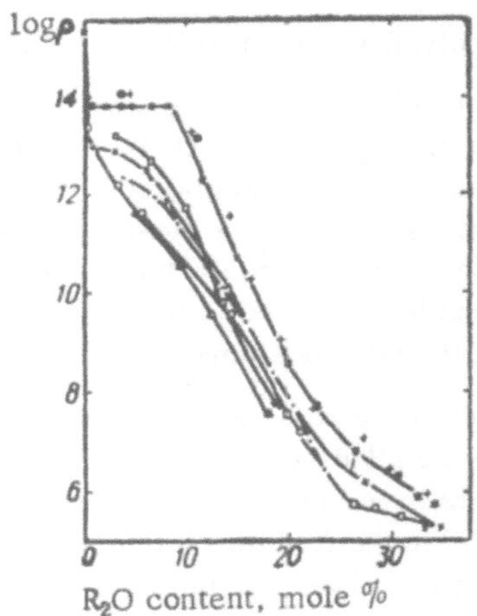

Fig. 14. Resistivity of two-component alkali borate glasses at 250°C. ○) $Li_2O-B_2O_3$ [96]; ●) $Li_2O-B_2O_3$ [143]; ×) $Na_2O-B_2O_3$ [155]; •) $K_2O-B_2O_3$ [96]; △) $Ag_2O-B_2O_3$ [103]; □) $Tl_2O-B_2O_3$ [104]; +) $Rb_2O-B_2O_3$ [96, 99]; ⊙) $Cs_2O-B_2O_3$ [96].

On the whole, our results are in satisfactory agreement with the data of other investigators (see [76]). However, the data for the system Li_2O-SiO_2 in a recent paper by Kuznetsov [64] differ from ours by two to three orders of magnitude. We have carried out, jointly with E. K. Mazurina, a very careful check of the results; this confirmed the earlier values.*

Systems $R_2O-B_2O_3$. The conductivity data available for this group of systems are the most reliable and complete of all the results for two-component glasses. There are two reasons for this. First, the whole group of systems was investigated in detail by the same school of workers, that of Myuller, by the same very reliable technique; second, it is easily possible to obtain borate glasses containing any desired small amounts of a second component. Figure 14 shows conductivity data for these systems at 250°C, which is the lowest temperature for which numerical values are reported.

It is evident from Fig. 14 that the relationships for all the systems are similar. All the oxides studied can be divided into two groups. In the first group, with K, Rb, and Cs ions, the resistivity is almost independent of composition in the range from 1 to 8% R_2O (this conclusion is tentative for Cs_2O because of the small number of points). This is followed by a sharp and approximately linear decrease of log ρ with increasing R_2O content up to 25% R_2O.

In the second group, which includes all the other systems, the decrease of log ρ becomes linear with increasing R_2O content from about 2% R_2O on the average.

*Recent data published by Charles [268] on the conductivity of several lithium silicate glasses are in good agreement with ours.

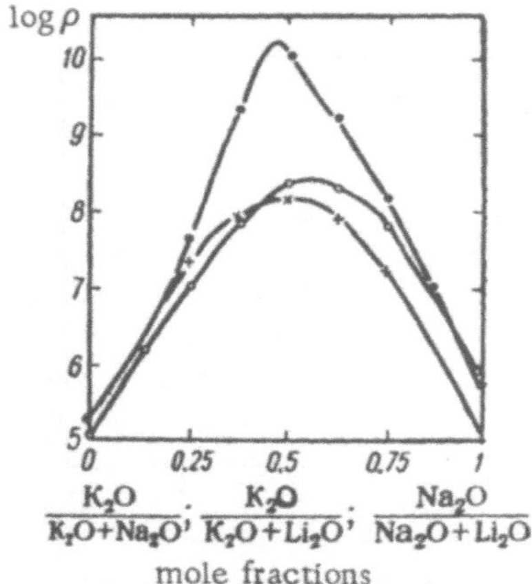

Fig. 15. Polyalkali effect at 150°C in alkali silicate glasses containing 33.3 mole % R_2O [76]. ○) $Na_2O-K_2O-SiO_2$; ×) $Li_2O-Na_2O-SiO_2$; ●) $Li_2O-K_2O-SiO_2$.

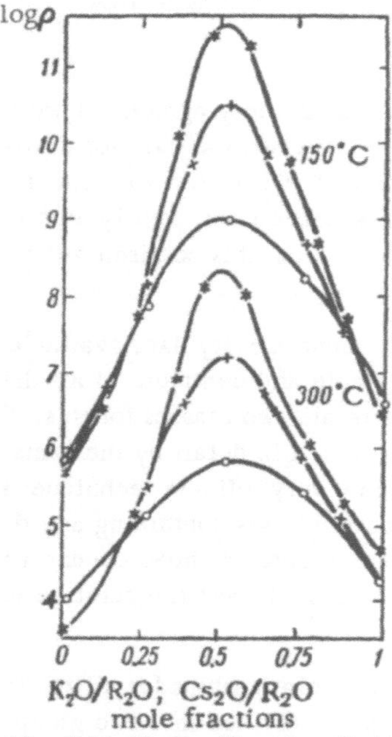

Fig. 16. Polyalkali effect in alkali silicate glasses containing 27 mole % R_2O [130]. ○) $Na_2O-K_2O-SiO_2$; +) $Li_2O-K_2O-SiO_2$; *) $Li_2O-Cs_2O-SiO_2$.

As in the case of silicate glasses, borate glasses containing potash have considerably higher resistivities than soda and lithium glasses. It is difficult to correlate the differences between the resistivities of the glasses within these groups with the size or other characteristics of the ions present in them.

Comparison of the conductivities of silicate and borate glasses shows that the first additions of R_2O have incomparably more effect on silica glass than on B_2O_3. As a result, in the region of moderate R_2O contents silicate glasses have much lower resistivities than borate glasses.

The Polyalkali Effect. When one alkali-metal oxide is replaced by another in a high-alkali glass, the resistance rises sharply. This important effect was discovered by Gehlhoff and Thomas [171] and studied in detail by a number of investigators, mainly in the Soviet Union [9, 73, 102, 105, 195, and others]. Skanavi proposed the term "neutralization effect" to describe this phenomenon (this refers to neutralization of the harmful effect of one alkali by introduction of another). This term cannot be regarded as entirely apt, because "neutralization" in scientific literature usually implies chemical neutralization. In accordance with R. L. Myuller's suggestion, we will describe the effect as the polyalkali effect.

Let us first consider the polyalkali effect in silicate glasses.

Figure 15 shows the effects of mutual replacement of the three main alkali oxides occurring in glasses: Li_2O, Na_2O, and K_2O. The total alkali oxide content for all the glasses is the same, 33.3%. The nature of the relation is quite clear from the figure. The considerably sharper maximum characteristic of the lithium–potassium series can be attributed to the large difference between the sizes of the two alkali-metal ions.

The effect can be expected to increase with increasing difference between the sizes of the alkali-metal ions. In fact (Fig. 16), the maximum becomes even more pronounced as we pass from the lithium–potassium to the lithium–cesium system. It is noteworthy that while rise of temperature reduces the maximum for the sodium–potassium system considerably, its influence on the polyalkali effect is much weaker for the lithium–potassium and especially the lithium–cesium system.

The results for the pseudoternary system consisting of the bisilicates of lithium, sodium, and potassium (Fig. 17) are interesting. The resistivity maximum is within the triangular diagram, nearer the lithium–potassium side. The maximum shifts toward binary lithium–potassium glasses with rise of temperature.

Let us now consider the influence of the total concentration of alkali oxides on the polyalkali effect. The relations are revealed most clearly if we compare series based on the fraction of alkali oxide replaced $R_2'O/\Sigma R_2O$) rather than on the percentage replacement [73, 76]. The magnitude and character of the effect are then seen to remain virtually constant in both the lithium–potassium and sodium–potassium series with total alkali oxide contents from 40 to 27%. Figure 18 represents the effect of varying the total alkali oxide content over a wider range on the polyalkali effect. The sodium–potassium system, for which the most data are

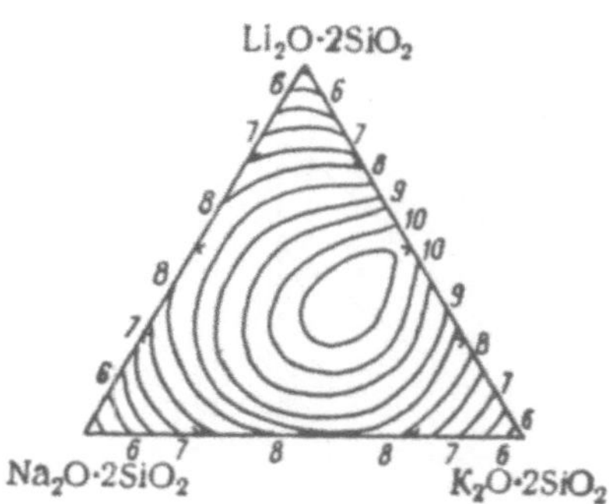

Fig. 17. Lines of equal log resistivity (isoresists) at 150°C [76].

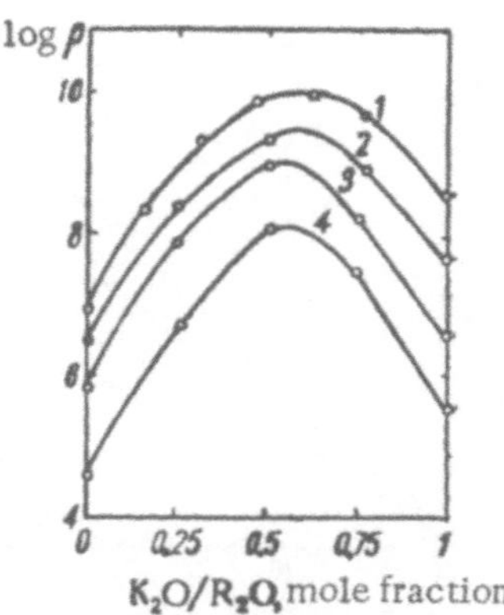

Fig. 18. Polyalkali effect in sodium–potassium silicate glasses at 150°C. Alkali oxide contents (mole %): 1) 13; 2) 20; 3) 27; 4) 40. Data taken from [76, 95, 135].

available, is taken as the example. It is seen that when the R_2O content is decreased below 27% the nature of the curves alters somewhat. This is not surprising because, as was noted above, the resistivity of potash glasses increases much more than that of soda glasses with decrease of R_2O content. Therefore, the maximum on the curve shifts toward pure potash glasses with decrease of R_2O, and the polyalkali effect begins to diminish on the potash side while remaining almost unchanged on the soda side. Further decrease of the R_2O content should result in a progressive displacement of the maximum toward the potash side, with the possibility that the effect may disappear entirely.

It is noteworthy that when the calculations are based on percentage contents the effect increases continuously with decrease of R_2O content down to 13%. In fact, in order to increase the resistance of a pure soda glass by 3 orders of magnitude 16% of Na_2O must be replaced by potassium oxide in a glass with 40% R_2O, whereas in a glass containing 13% R_2O the same result is achieved by replacement of only 6% Na_2O.

To judge by the available data [6, 76, 101], the characteristic relationships for the polyalkali effect in borate and borosilicate glasses are quite analogous to those found for silicate glasses. Curves characterizing this effect in borate, borosilicate, and silicate glasses with the same alkali oxide content are remarkably similar (Fig. 19).* With regard to the influence of decrease of the alkali oxide content on the effect in borate glasses, Markin's data [98] indicate that in the lithium–sodium system the polyalkali effect alters only slightly down to R_2O contents of 12-13%, while in the lithium–potassium system it decreases, disappearing almost entirely at 12-13% R_2O. It follows that no definite regularities can be detected at present.

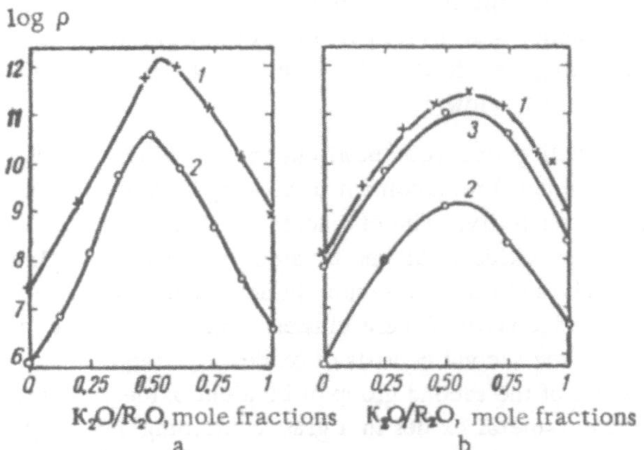

Fig. 19. Comparison of the polyalkali effect at 150°C. 1) Borate glasses [98]; 2) silicate glasses [76]; 3) borosilicate glasses [101]. a) Lithium–potassium glasses; b) sodium–potassium glasses.

Influence of Bivalent-Metal Oxides. This was first studied by Gehlhoff and Thomas [171]. Their investigation contains serious errors, apparently associated with incorrect calculations of the batches for some of the glass compositions (see p. 45). The results of a series of systematic investigations in this field are now avail-

*Ivanov [256] showed that the same applied to the polyalkali effect in germanate glasses.

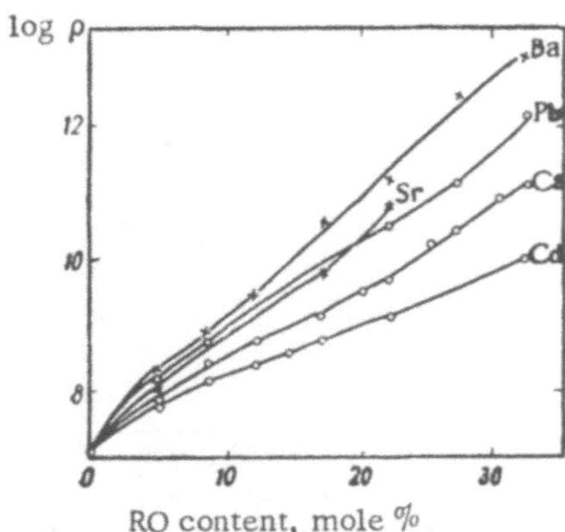

Fig. 20. Effect on the resistivity at 150°C of replacement of silica by bivalent-metal oxides in glass containing 13 mole % Na_2O. Data of Mazurin and V. S. Molchanov.

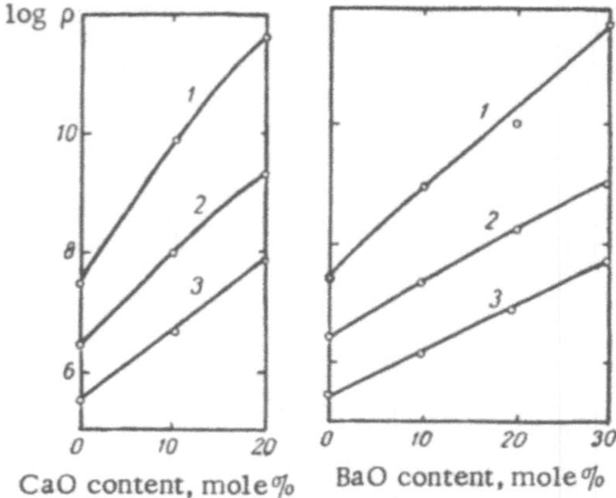

Fig. 21. Effect on the resistivity at 150°C of replacement of silica by bivalent oxides in sodium silicate glasses. Sodium oxide contents (mole %): 1) 10; 2) 20; 3) 30. Data of Mazurin and M. I. Kalinin.

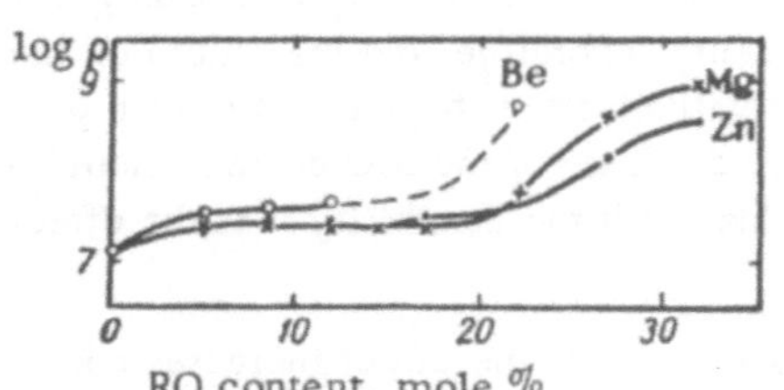

Fig. 22. As in Fig. 20.

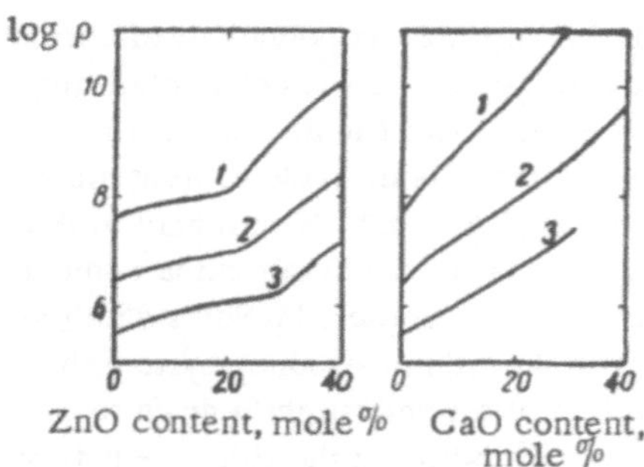

Fig. 23. As in Fig. 21.

able (the following systems have been studied: $Na_2O-BaO-SiO_2$ [9, 207]; $Na_2O-PbO-SiO_2$ [61], and $Na_2O-RO-SiO_2$ [84, 85]).

The results obtained by Mazurin and his co-workers are discussed below. They are generally in good agreement with the data reported by other investigators, and differ considerably only from the results of Gehlhoff and Thomas [171] and of Kuznetsov et al. [87].

The influence of bivalent oxides has been studied most fully in the case of sodium silicate glasses.

Conductivity of the system $Na_2O-RO-SiO_2$, with 10-30% Na_2O and 0-20% RO, was studied [84, 85]. The results were subsequently augmented by Mazurin and M. I. Kalinin, who studied certain compositions containing 30 and 40% RO, and by Mazurin and V. S. Molchanov, who investigated the effect of added RO in glasses containing 13% Na_2O. The results of these investigations are summarized below.*

In high-alkali glasses (compositions with less than 10% R_2O are not considered here) replacement of SiO_2 by bivalent-metal oxides increases resistivity. The character of this increase differs sharply for different oxides. All the bivalent oxides used in glassmaking can be divided into two groups in accordance with their effects on the conductivity of these glasses. The first group comprises oxides containing relatively small ions: Be, Mg, and Zn; the second consists of oxides with larger ions, namely Cd, Ca, Sr, Pb, and Ba. We first examine the influence of the second group of bivalent oxides. Figure 20 represents the results of replacement of SiO_2 by these bivalent-metal oxides in a glass containing 13% Na_2O. For all the systems studied, log ρ increases linearly in the first approximation with increase of RO to 32%. It is clearly seen that the effects of the oxides in raising the resistivity of glass increase with increasing radius of the bivalent ion.

*Similar results were obtained by Lengyel and Boksay [270]. However, direct comparison of the data is difficult because these workers replaced Na_2O by RO, i.e., altered the content of charge carriers.

TABLE 3. Effect of Partial Replacement of R'O by R"O on Resistivity of Alkali Glasses* [Glass composition (mole %): Na_2O, 17; RO, 28; SiO_2, 55]

Ro content, mole %				$\log \rho$ 150° C	$\Delta \log \rho$† 150° C	$\log \rho$ 300° C	$\Delta \log \rho$† 300° C
MgO	ZnO	CaO	BaO				
28	—	—	—	7.49	—	5.15	—
—	28	—	—	7.49	—	5.15	—
—	—	28	—	9.65	—	6.68	—
—	—	—	28	10.91	—	7.54	—
14	14	—	—	7.75	0.26	5.28	0.13
14	—	14	—	8.97	0.40	6.22	0.31
14	—	—	14	9.87	0.67	6.83	0.48
—	14	14	—	9.03	0.46	6.23	0.32
—	—	14	14	10.60	0.32	7.38	0.27
—	14	—	14	9.62	0.42	6.66	0.31
9.3	9.3	—	9.3	8.95	0.32	6.08	0.13
9.3	9.3	9.3	—	9.77	0.56	6.05	0.39
7	7	7	7	9.32	0.34	6.37	0.24

*Data of Mazurin and V. S. Molchanov.

†$\Delta \log \rho = \log \rho_{exptl} - \log \rho_{calc}$, where $\log \rho_{calc}$ is calculated by the additivity rule from data on three-component glasses.

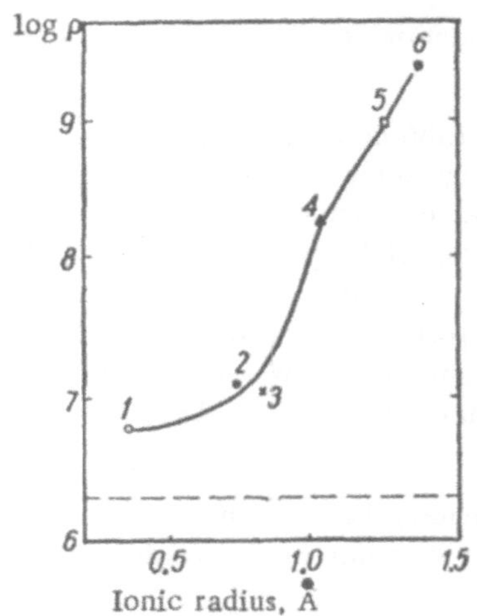

Fig. 24. Resistivities at 150°C of glasses containing 8.5 M sodium oxide and 8.5 M of various bivalent oxides: 1) Be; 2) Mg; 3) Zn; 4) Ca; 5) Pb; 6) Ba. The dashed line corresponds to glass without RO.

The effects of these bivalent oxides remain of the same character when the Na_2O content is varied in the range of 10-30%. At the same time, the influence of the bivalent ion diminishes with increasing sodium oxide content in all the systems (data on the calcium and barium systems are given as an example in Fig. 21). Therefore decrease of the sodium oxide content produces a substantially greater increase of resistivity in glasses containing considerable amounts of bivalent oxides of the second group than in glasses without RO.

The behavior of bivalent oxides of the first group is more complex. Once again it can be seen most clearly in series of glasses containing 13% Na_2O (Fig. 22). Replacement of SiO_2 by the oxides of Be, Mg, or Zn in amounts up to 20% has very little effect on the resistivity of the glass (there is only a certain increase of resistivity resulting from the first replacements). The curve then bends sharply and the resistivity increases considerably with further substitution up to 30%. Increase of the Na_2O content does not alter the general character of the curve (Fig. 23) but the bend is less sharp and is displaced toward higher RO contents. Thus, here again the influence of other components becomes greater with decrease of the Na_2O content.

Data on the two groups of oxides can be combined in a plot of resistivity versus the ionic radius of the bivalent ion in the glass. Figure 24 shows that such a plot gives a smooth curve.

The combined influence of several bivalent oxides on conductivity is a problem of practical importance. Mazurin and V. S. Molchanov carried out a special investigation in this field. Glasses of the total composition 17% Na_2O, 28% RO, 55% SiO_2 were made for the purpose. These glasses contained bivalent oxides either singly or jointly in equal amounts. Data on their resistivities are presented in Table 3. It was found that if the resistivities of the complex glasses were calculated from data for simple glasses on the assumption that $\log \rho$ varies additively with the composition, the experimental results are always higher than the calculated values.

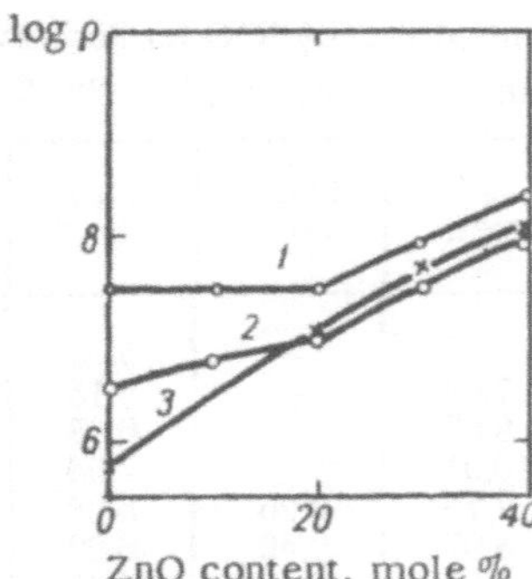

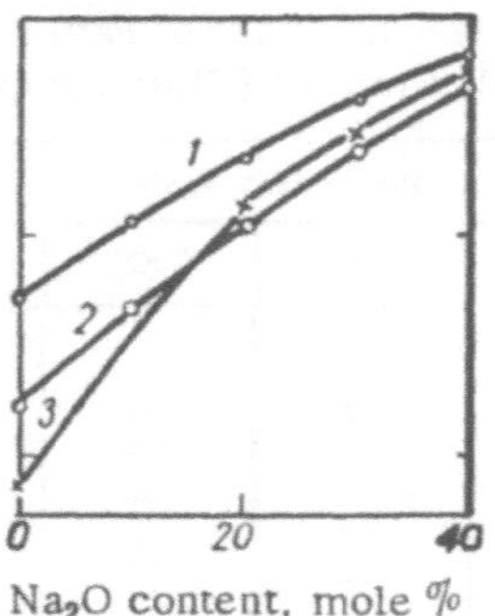

Fig. 25. Effect of replacement of silica by bivalent-metal oxides on the resistivity at 150°C of alkali silicate glasses containing 20 mole % R_2O. 1) Potassium glasses; 2) sodium glasses; 3) lithium glasses. Data of Mazurin and M. I. Kalinin.

The table gives values of $\Delta \log \rho$ at 150°C, where $\Delta \log \rho = \log \rho_{exptl} - \log \rho_{calc}$. It is seen that the difference increases with increasing difference between the sizes of the bivalent ions in the glass (it is necessary to take into account the experimental variance of $\Delta \log \rho$, which is approximately double the variance of $\log \rho$ values). However, for most of the compositions $\Delta \log \rho$ is in the range 0.3-0.5. Similar results were obtained for series of glasses of the total compositions 13% Na_2O, 27% RO, 60% SiO_2 and 12% Na_2O, 12% RO, 76% SiO_2. A similar result for the glass series 16% Na_2O, 16% RO, 68% SiO_2, was recently obtained by Lengyel and Boksay [199].

The next, and a very significant, question is how the influence of bivalent oxides changes when different compositions of the original glasses are used.

First we investigated the effects of bivalent oxides on the resistivity of lithium and potassium silicate glasses (Fig. 25). Two bivalent oxides were studied: zinc oxide, which belongs to the first group of bivalent oxides by the nature of its influence on conductivity, and calcium oxide, a representative of the second group. The figure shows that the transition from lithium to potassium glasses has an especially strong influence on the behavior of zinc oxide. The increase of resistivity due to introduction of zinc oxide is greater in lithium than in potassium glasses, and the bend of the curve which is characteristic of the first group of bivalent ions is more pronounced for potassium glasses while probably not appearing at all for lithium glasses. We do not have sufficient experimental data for a more definite conclusion (as is well known, many lithium glasses have an exceptionally high tendency to crystallization).

The influence of calcium oxide on resistivity was found to be approximately the same in all three series. However, there is a fairly noticeable tendency for its effect to increase in the transition from potassium to lithium glasses.

TABLE 4. Changes of Resistivity ($\Delta \log \rho$) of Glasses of the Composition 17 mole % R_2O and 83 mole % SiO_2 on Replacement of 13 mole % SiO_2 by Various Bivalent-Metal Oxides [95]

RO	$\Delta \log \rho$ at 150°C	
	Sodium glasses	Potassium glasses
MgO	+0.77	−0.08
ZnO	+0.55	+0.05
CaO	+1.72	+1.18
BaO	+2.41	+2.26
PbO	+2.53	+2.57

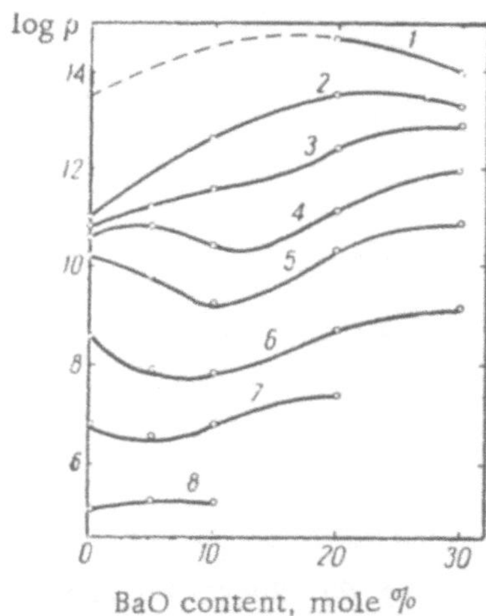

Fig. 26. Effect of replacement of boric anhydride by barium oxide on the resistivity of sodium glasses at 300°C. Sodium oxide contents (mole %): 1) 0; 2) 2.5; 3) 5; 4) 7.5; 5) 10; 6) 15; 7) 20; 8) 30. Data of Mazurin, Sheng T'ing-k'un, and V. I. Trefilova.

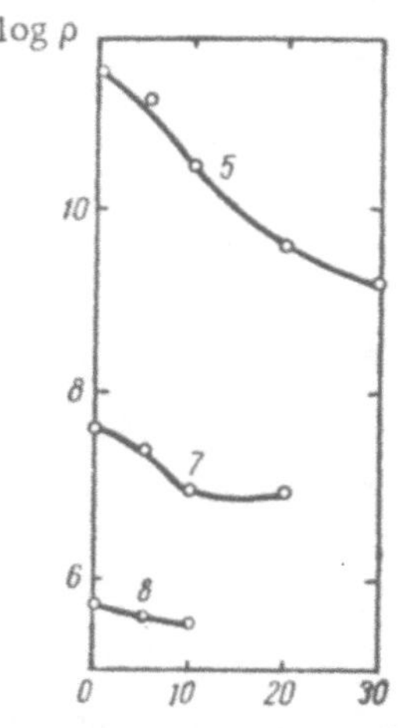

Fig. 27. Analogous to Fig. 26 (data of Mazurin and Sheng T'ing-k'un).

The effects of certain bivalent oxides on the resistivity of sodium and potassium glasses are compared in Table 4. The small amount of experimental data in this table appears to confirm the conclusions drawn from the data of Fig. 25.

Mazurin, Sheng T'ing-k'un, and V. I. Trefilova investigated the effects of bivalent oxides on the resistivity of alkali borate glasses. It was shown earlier in this article that the polyalkali effect is exactly the same in silicate as in borate glasses. It could therefore be supposed that a change in the nature of the glass-forming oxide should not result in any essential change in the influence of bivalent oxides on resistivity. However, the result of the investigation was entirely unexpected (Figs. 26 and 27). The influence of bivalent oxides on the resistivity of alkali borate glasses depends to a very great extent on the alkali-metal oxide content of the original glass. At low Na_2O concentrations (up to approximately 6%) the increase of resistivity produced by introduction of BaO is greater at lower alkali oxide contents. The nature of the relation gradually changes with increasing Na_2O content, and with 10% Na_2O a distinct minimum appears on the curve, corresponding to approximately 10% BaO. On further increase of the Na_2O content to 20 and especially to 30% the resistivity minimum shifts toward lower BaO contents and becomes progressively less distinct; finally, at 30%, replacement of B_2O_3 by barium oxide ceases to influence the resistivity of the glass.

The earlier data of Markin and Myuller [97, 98] for the sodium–barium borate system do not, in general, contradict ours, although a complete comparison is impossible because the series were designed quite differently.

The effects of calcium oxide and barium oxide are quite similar. However, introduction of magnesium oxide, even in considerable amounts, results in a continuous fall of resistivity of glasses containing 10-30% Na_2O. Here again the influence of the replacement becomes gradually weaker for more alkaline compositions.

The bivalent oxides can probably be divided into the same groups in accordance with their influence on the resistivities of both alkali borate and alkali silicate glasses.

Replacement of B_2O_3 by barium oxide produces the same effect in lithium and potassium glasses as in sodium glasses (Fig. 28).

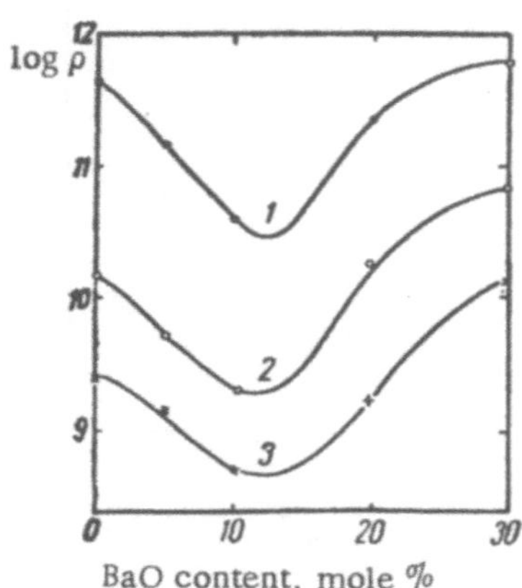

Fig. 28. Effect on resistivity at 300°C of replacement of boric anhydride by barium oxide in glasses containing 10% alkali oxides. 1) Potassium glasses; 2) sodium glasses; 3) lithium glasses. Data of Mazurin and Trefilova.

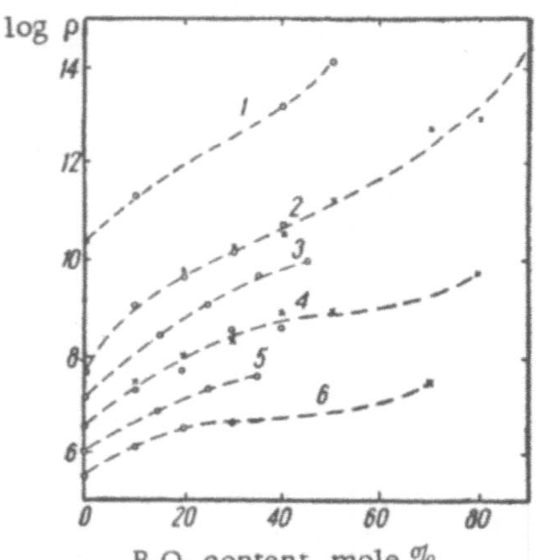

Fig. 29. Effect of replacement of silica by boric anhydride on the resistivity of sodium glasses at 150°C. O) Mazurin's data; x) data of Mazurin, V. I. Gutman, and M. I. Smirnova. Sodium oxide contents (mole %): 1) 5; 2) 10; 3) 15; 4) 20; 5) 25; 6) 30.

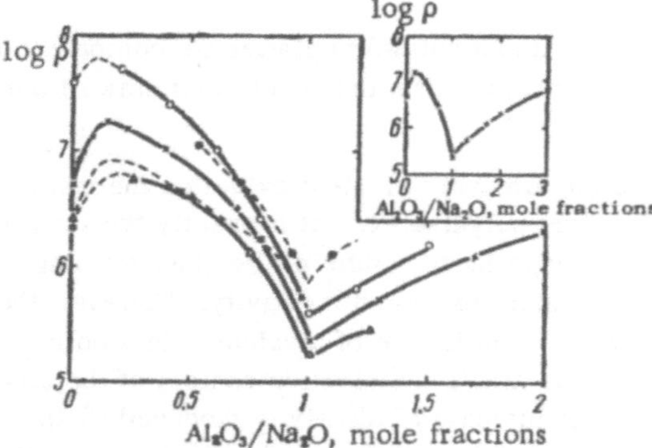

Fig. 30. Effect of replacement of silica by alumina on the resistivity of sodium silicate glasses at 150°C. Top right shows the same on a reduced scale. x) 13 mole % Na_2O (data of V. A. Tsekhomskii and Mazurin); •) 20 mole % Na_2O [49]; ●) 16.5 mole % Na_2O [49]; O) 20 mole % Na_2O [185]; Δ) 20 mole % Na_2O (data of V. A. Tsekhomskii and Mazurin).

Influence of Simultaneous Presence of Tri- and Quadrivalent Oxides. The effects of partial replacement of silica in silicate glasses by boric anhydride and alumina are of the most interest here.

Figure 29 shows the results of the author's experiments with glasses kindly supplied by A. A. Kefeli, and also the results reported by Mazurin, V. I. Gutman, and M. I. Smirnova. As the composition of glasses in the system $Na_2O-B_2O_3-SiO_2$ can alter considerably during the melting, glasses for which analytical data were available were used in the first investigation. In the second, some selective analyses were performed. Despite the excellent agreement between the analytical data and the composition by synthesis (within 1% for most glasses), our conductivity determinations for certain series of glasses of the same composition, but melted by a somewhat different procedure (and also determinations carried out by Appen and Kan Fu-hsi [3]), showed considerable discrepancies from the data of Fig. 29 (as much as 0.3-0.4 log ρ), much greater than the variance of our results for any other glasses. This is undoubtedly due to the high sensitivity of the structure of sodium borosilicate glasses to their thermal history. We therefore regard the data in Fig. 29 as approximate and have indicated the curves by dashed lines. However, the curves represent the general nature of the relationships fairly reliably.

There are no grounds for believing that the nature of the relationship for potassium borosilicate glasses would be different in principle from that found for sodium borosilicate glasses. In fact, Markin's data [101] appear to indicate that the electrical properties of sodium and potassium borosilicate glasses are similar.

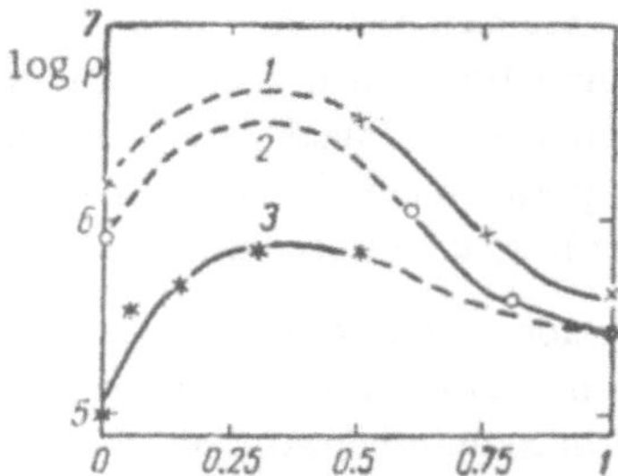

Al_2O_3/Li_2O, mole fractions

Fig. 31. Influence of alumina on resistivity of lithium silicate glasses at 150°C. 1) Replacement of SiO_2 by alumina in glass containing 25 mole % Li_2O; 2) the same in glass containing 20 mole % Li_2O; 3) addition of Al_2O_3 to glass containing 33.3 mole % Li_2O. Data of S. K. Dubrovo, Yu. A. Shmidt, and Mazurin.

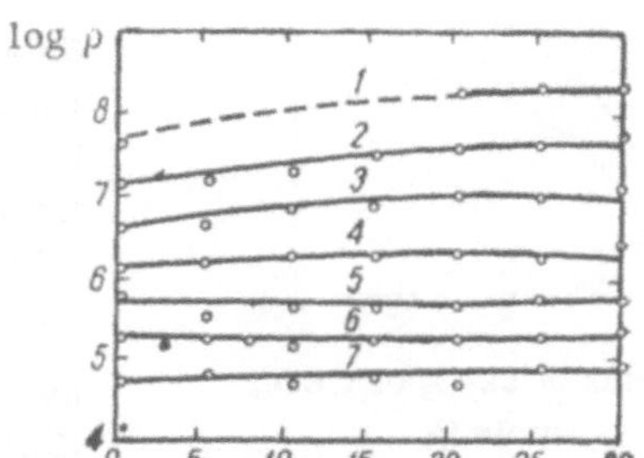

TiO_2 content, % by weight

Fig. 32. Effect of replacement of silica by titanium dioxide [24] at 150°C. Sodium oxide contents (% by weight): 1) 10; 2) 15; 3) 20; 4) 25; 5) 30; 6) 35; 7) 40.

The effect of replacement of silica by alumina in soda glasses has been the subject of some recent investigations [49, 185]. Figure 30 is based on literature data and on results obtained by the author with V. A. Tsekhomskii.* The curves show that the relationship remains roughly the same at all Na_2O contents. The resistivity increases appreciably during replacement of the first few percent; it then decreases considerably and reaches a minimum at the ratio $Na_2O/Al_2O_3 = 1$. Further replacement results in increased resistivity.† The absence of the initial maximum on the curves obtained by Ioffe and Khvostenko [49] is evidently attributable to lack of data on glasses with low alumina contents.

The curves in Fig. 30 lead to the conclusion that the influence of alumina increases with decreasing sodium oxide content.

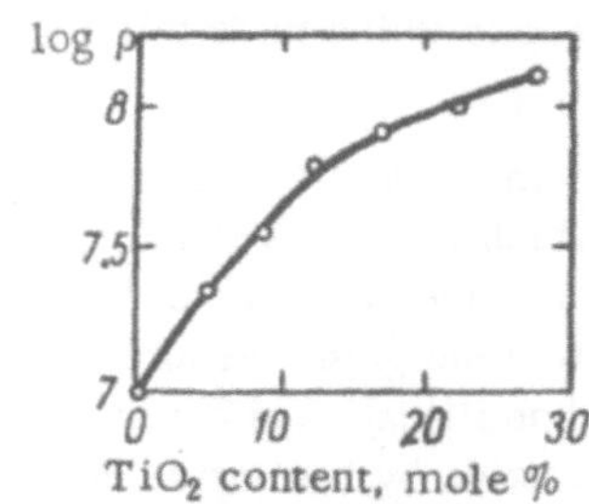

TiO_2 content, mole %

Fig. 33. Effect of replacement of silica by titanium dioxide at 150°C. Na_2O content 13 mole %. Data of Mazurin and V. S. Molchanov.

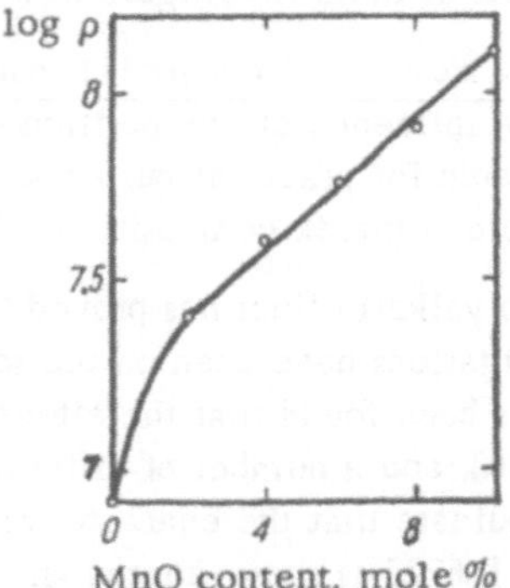

MnO content, mole %

Fig. 34. Effect of replacement of silica by manganese oxide at 150°C. Na_2O content 18 mole % [77].

*Numerical data on this work are partially published in [266].

†The curve characterizing replacement of SiO_2 by gallium oxide in silica glass [257] is exactly similar in character.

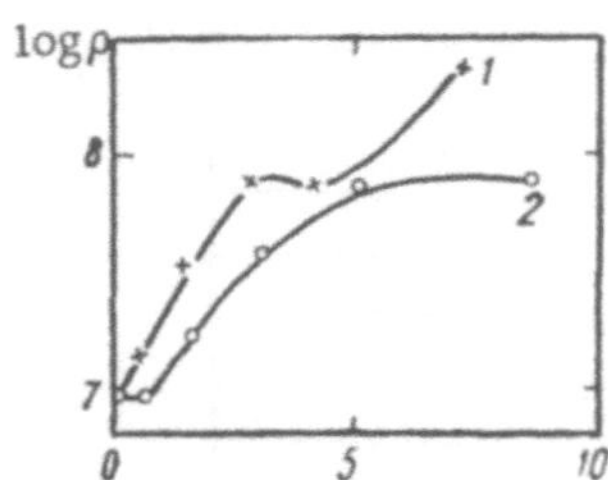

Contents of La_2O_3 or CeO_2, mole %

Fig. 35. Effect of replacement of silica by rare-earth oxides on resistivity at 150°C. Na_2O content 13 mole %. 1) Replacement by lanthanum oxide; 2) replacement by cerium oxide. Data of Mazurin and V. S. Molchanov.

The information available at present on the system $Li_2O-Al_2O_3-SiO_2$ is limited (Fig. 31). These data indicate fairly clearly that alumina has the same effect in lithium silicate as in sodium silicate glasses. It can also probably be concluded from these results that the first additions of alumina in lithium glasses increase the resistivity more, and subsequent additions lower it less, than in sodium glasses.

The available information on the joint influence of boric anhydride and alumina is confined to the work of Appen and Kan Fu-hsi [3]. Their results show that in sodium borosilicate glasses replacement of silica by alumina, starting at 4% (lower percentages were not investigated), lowers the resistivity. This effect is intensified with increasing boric anhydride and decreasing sodium oxide contents. The Na_2O content has a greater influence than the B_2O_3 content in this respect.

The resistivity of the system $Na_2O-TiO_2-SiO_2$ (Fig. 32) was studied in detail by Vargin and Antonova [24]. Replacement of silica by titanium dioxide raises the resistivity slightly; the effect is less at higher alkali oxide contents in the original glass. However, in assessment of the results it should be remembered that the replacement was in percentages by weight; replacement by molecular percentages would intensify the effect considerably. The results of molecular replacement of SiO_2 by titanium dioxide in glasses containing 13% Na_2O are plotted in Fig. 33.

Effects of Other Components. Introduction of fluorine raises the resistivity of simple high-alkali glasses [75]. Manganese oxide raises the resistivity of high-alkali glasses (Fig. 34) considerably [77].

Replacement of SiO_2 by certain rare-earth oxides raises the resistivity; the increase is especially pronounced with La_2O_3 (Fig. 35). Replacement of SiO_2 by niobium pentoxide has little influence on the conductivity of alkali silicate glasses [269].

The only results available for other oxides were obtained in investigations in which some very diverse oxides were introduced by replacement or addition into an original glass of definite composition [78, 158, 188]. This method of study is very useful for investigating properties which vary, at least in the first approximation, additively with the composition of the glass. However, in some cases the influence of composition on conductivity of glasses deviates from additivity (for example, when one alkali oxide is replaced by another). Therefore the results of these investigations are not discussed here.

Influence of Composition on Resistivity of Multicomponent Glasses. We have examined the influence of composition on resistivity of simple high-alkali glasses. In order to make the material applicable for practical purposes, we must determine the extent to which these relationships apply to glasses of more complex composition.

The polyalkali effect has proved to be very persistent in relation to variations of glass composition. Most of the investigations have been on the sodium–potassium system, which is the one of most practical significance. It has been found that the effect is almost unchanged in magnitude or character by introduction of MnO [77], TiO_2 [80], and a number of different bivalent oxides [79, 95] into the glass. As already noted, Markin's data [101] indicate that the character of the effect is the same for borosilicate and borate as for silicate glasses. Shumitskaya [152] has shown that progressive replacement of barium oxide by lead oxide in borate and silicate glasses also has little influence on the effect. Some results from [95] are given in Fig. 36 as an example.

According to the results of Mazurin, Golikova, and Shtol'tser [79] and of Mazurin, T. A. Polyakova, and L. A. Kabanova, the relationship also retains its character for lithium–potassium glasses of increasingly complex composition.

A very important conclusion can therefore be drawn: the polyalkali effect in complex glasses depends mainly on the amount and proportion of alkali oxides in the glass and varies little under the influence of other oxides introduced into the glass. Therefore we can assume that the influence of the polyalkali effect on the logarithm of resistivity obeys the additivity rule.

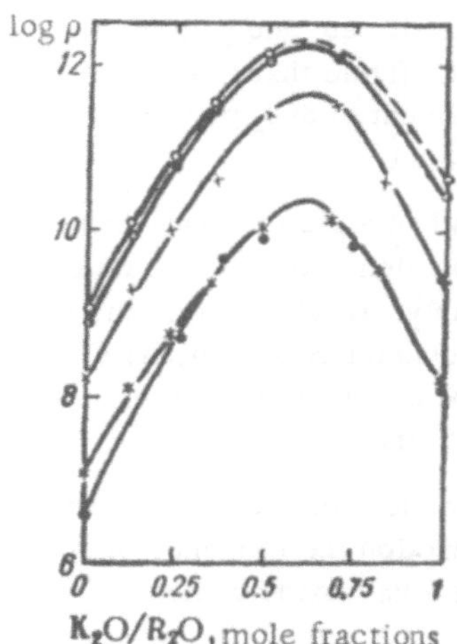

Fig. 36. Effect of replacement of sodium oxide by potassium oxide on the resistivity at 150°C of glasses of the general composition (mole %): R_2O, 17; RO, 13; SiO_2, 70, where RO is: ○) PbO; ⊙) BaO; x) CaO; •) ZnO. ●) Series of glasses of the composition 16% R_2O, 84% SiO_2 [95].

Of course, even for a particular case, additivity of a property such as resistivity is only a very relative concept and cannot in any way compare in accuracy with additivity of refraction or molecular volume. In fact, as was pointed out above, bivalent oxides differ in their influence on the conductivity of sodium and potassium glasses, and this alone makes complete additivity impossible. Several other factors also play a part. However, even approximate additivity is of very great importance for practical utilization of data on simple systems. A good example, which confirms, in particular, the value of data on the polyalkali effect for improving the properties of complex glasses, is provided by Figs. 37 and 38, which represent results obtained in a study of the polyalkali effect in a complex glass corresponding in composition to the known No. 2 vacuum tube glass. In the series of glasses investigated in this study the total molecular R_2O content was kept constant with mutual replacement of the oxides of Li, Na, and K.

In comparison with the corresponding triangular diagram for glasses of the composition 33.3% R_2O, 66.6% SiO_2 (see Fig. 17), the resistivity maximum is shifted toward the potassium corner; this is due, as was pointed out on p. 23,to a decrease of the total amount of alkali. The relationship remains strictly of the same character in all other respects. It is noteworthy that a considerable group of compositions lies in the region where $t_{\varkappa-100}$ exceeds 350°C; i.e., $t_{\varkappa-100}$ of these glasses is comparable to the $t_{\varkappa-100}$ of such glasses as ZS-9 and No. 40 (containing 2-3% R_2O) and is higher than the $t_{\varkappa-100}$ of all other glasses used in our vacuum tube industry [70].

The converse conclusion can also be drawn from all the material considered above: the influence of bivalent oxides in glasses containing several univalent oxides does not differ greatly from their influence in glasses containing single alkali oxides. Apparently replacement of small amounts (up to 10%) of silica by the oxides of boron or aluminum also has little significant effect on the general influence of bivalent oxides on the resistivity of alkali glasses. However, a complete change in the nature of the relationship is to be expected when considerable amounts of B_2O_3 are introduced into silicate glass. This view is supported, in particular, by the data of Mazurin and V. I. Trefilova (Fig. 39). The effect of CaO in borosilicate glass is intermediate between

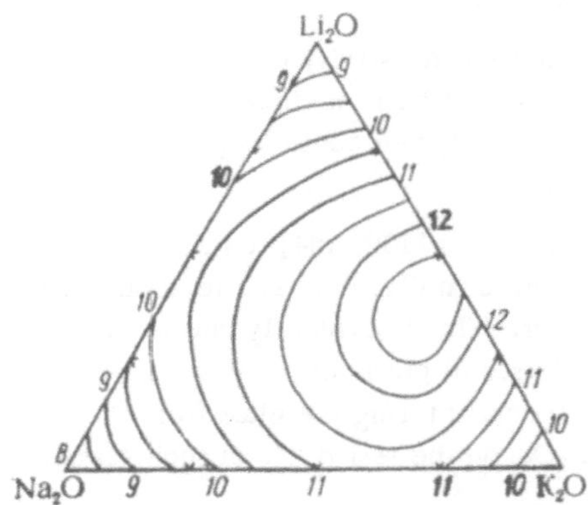

Fig. 37. Lines of equal log resistivity (isoresists) at 150°C for glasses of the composition (mole %): (Li_2O + Na_2O + K_2O), 16.2; BaO, 0.8; CaO, 5.9; MgO, 5.2; SiO_2, 71.9. Data of Mazurin and L. A. Kabanova.

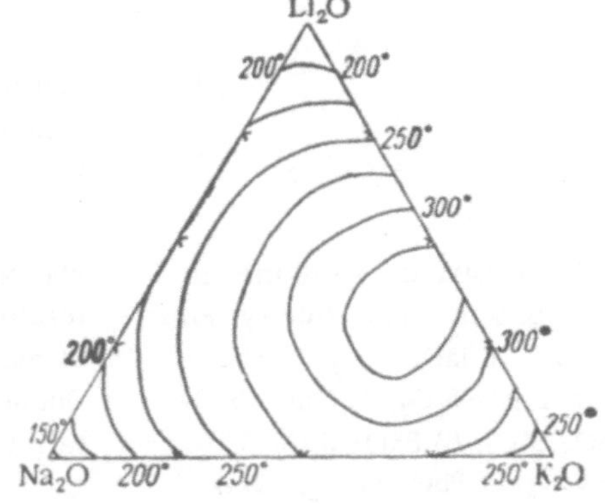

Fig. 38. Lines of equal $t_{\varkappa-100}$ for glasses of the system of Fig. 37. Data of Mazurin and L. A. Kabanova.

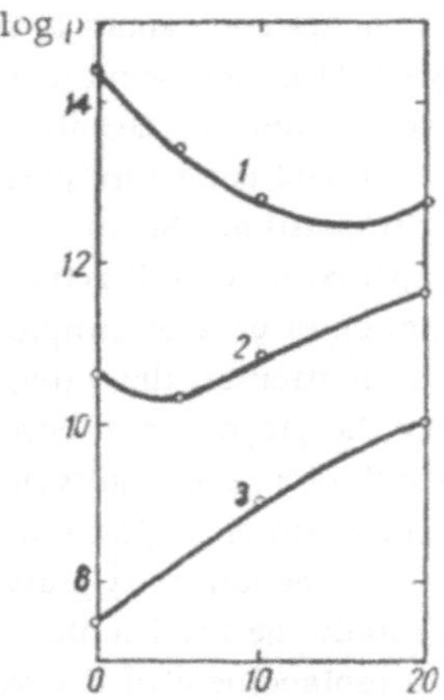

CaO content, mole %

Fig. 39. Effect on resistivity at 150°C of replacement of glass-forming oxides by calcium oxide in borate, borosilicate, and silicate glasses containing 10 mole % Na_2O (data on borate glasses were found by extrapolation from higher temperatures). 1) B_2O_3; 2) $SiO_2/B_2O_3 = 1/1$; 3) SiO_2.

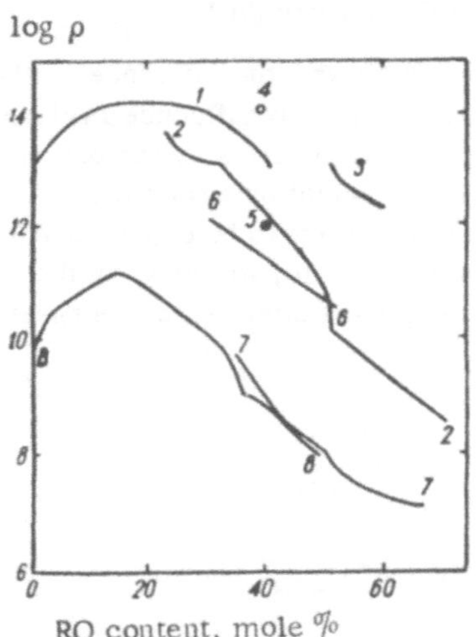

RO content, mole %

Fig. 40. Resistivities of glasses in two-component alkali-free systems at 300°C. 1) $BaO-B_2O_3$; 2) $PbO-B_2O_3$; 3) $ZnO-B_2O_3$; 4) $CaO-B_2O_3$; 5) $CaO-SiO_2$; 6) $BaO-SiO_2$; 7) $PbO-SiO_2$; 8) $PbO-GeO_2$. Data of Mazurin and [33, 34, 35, 72, 97, 102].

its effects on the resistivities of pure silicate and pure borate glasses. It is also clear from the curves in the figure that in glasses containing considerable amounts of RO replacement of SiO_2 by boric anhydride has much less effect than in simpler glasses.

It is interesting to note that when a glass is made more complex in composition by introduction of bivalent oxides, the nature of the influence of alumina alters considerably. It follows from the work of Moore and De Silva [200] that the characteristic resistivity minimum with small amounts of Al_2O_3 disappears, and introduction of Al_2O_3 merely results in a steady decrease of resistivity.

It can also be concluded from the results discussed above that in the case of glasses with alkali conduction the influence of composition generally increases with decreasing alkali oxide content.

Glasses with Nonalkaline Conduction

It must be borne in mind that most of the relationships representing the influence of composition on conductivity of systems with nonalkaline conduction differ in principle from the corresponding relationships for alkali glasses. It should also be taken into account that many of the glasses included in the systems under consideration here are difficult to make. The determinations are also more complicated, because many of these glasses have very high resistivities. Therefore the discrepancies between the absolute resistivity values obtained by different investigators for alkali-free and low-alkali glasses are generally greater than the corresponding discrepancies for high-alkali systems.

Two-Component Systems. The results of various workers on two-component borate and silicate systems studied up to the present time are summarized in Fig. 40. The following main conclusions can be drawn from the plots. Borate glasses have higher resistivities than silicate glasses. When the bivalent oxide content is over 20-30%, further introduction of the same oxide leads to a considerable decrease of resistivity. Glasses with lead oxide have lower resistivities than the others. Apparently, in lead-free systems the resistivity increases with decreasing radius of the bivalent ion.

It is noteworthy that the resistivity of boric anhydride is raised considerably by introduction of barium oxide. This would probably also apply to silicate glasses if they could be prepared with low contents of bivalent oxides.

The results of two studies [132, 164] on the conductivity of simple phosphate glasses are plotted in Fig. 41. The reproducibility of the results is remarkably poor. This is evidently due to the impossibility of melting phosphate glasses in platinum vessels, and to the strong influence of other components entering the glass from the crucible material, in particular Al_2O_3, on the resistivity of such glasses.

Influence of Simultaneous Presence of SiO_2, B_2O_3, and Al_2O_3. Since nearly all alkali-free glasses used in practice contain alumina, the influence of this oxide on glass conductivity is of particular interest.

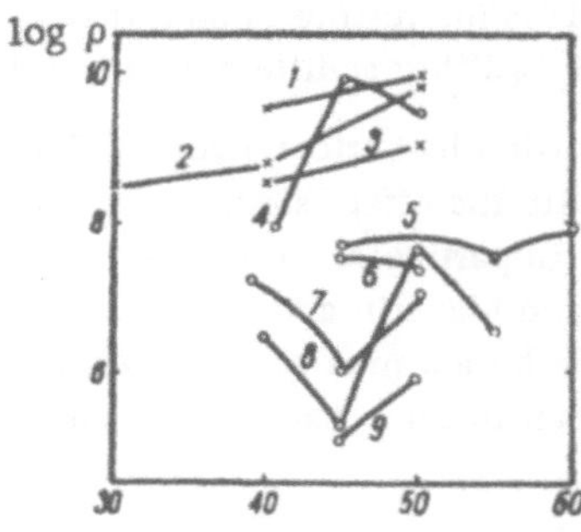

RO content, mole %

Fig. 41. Resistivities of two-component phosphate glasses at 300°C, from the data of Elyard [164] and Petrosyan [132]. 1) Mg; 2) Ca; 3) Ba; 4) Mg; 5) Zn; 6) Ca; 7) Sr; 8) Ba; 9) Pb. 1, 2, 3) Petrosyan's data; 4-9) Elyard's data.

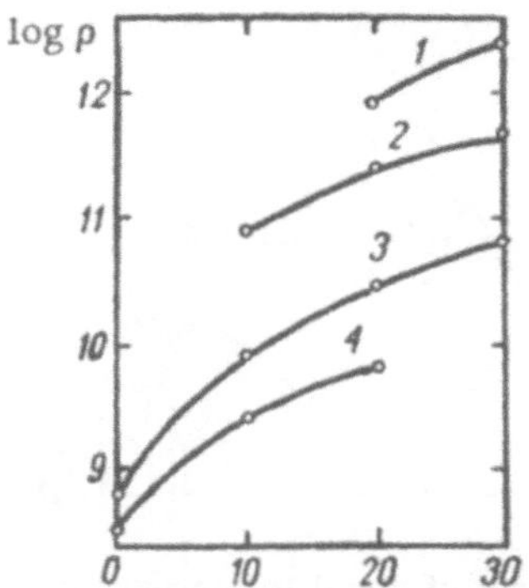

Al_2O_3 content, mole %

Fig. 42. Effect of replacement of silica by alumina on the resistivity of lead silicate glasses at 300°C [88]. PbO contents (mole %): 1) 30; 2) 40; 3) 50; 4) 60.

This subject has been studied in detail with the system $PbO-Al_2O_3-SiO_2$ by Mazurin and Brailovskii [88] (Fig. 42). The figure shows that at all contents of lead oxide and of Al_2O_3 introduction of more alumina raises the resistivity. The first portions of alumina have the greatest effect on the resistivity, and the influence then becomes weaker.

From a comparison of certain data taken from various literature sources, the authors put forward the suggestion (also advanced independently by Khar'yuzov [149]) that introduction of alumina raises the resistivity of the most diverse alkali-free silicate glasses, and not only that of lead silicate glasses.* This effect of alumina has now been established for barium and calcium as well as for lead glasses.

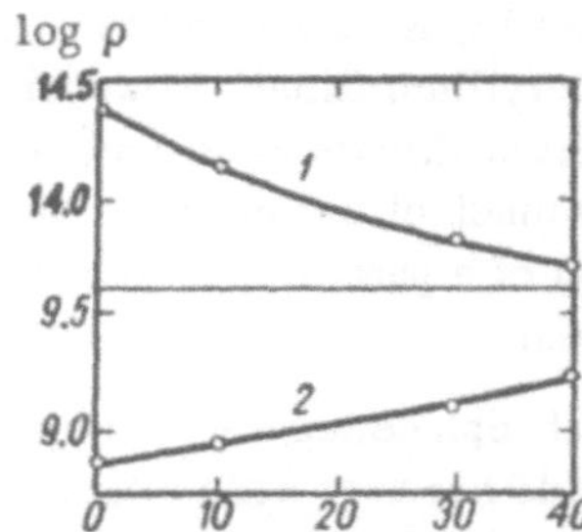

Al_2O_3 content, mole %

Fig. 43. Effect of replacement of B_2O_3 by alumina on the resistivity of calcium borate glass containing 40 mole % CaO at 300°C (1) and 600°C (2). Data of Mazurin and G. A. Pavlova.

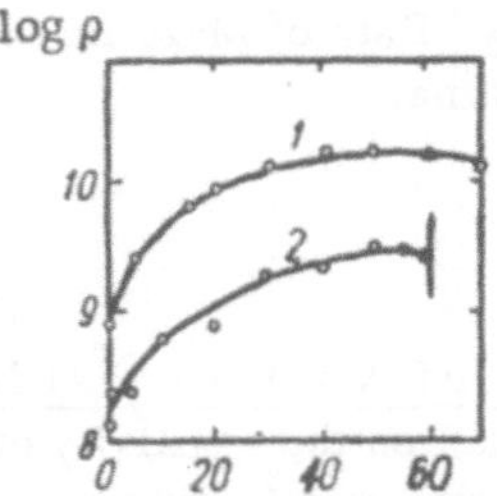

B_2O_3 content, mole %

Fig. 44. Effect of replacement of silica by boric anhydride on the resistivity of barium silicate glasses at 500°C [150]. BaO contents: 1) 30 mole %; 2) 40 mole %.

*The same relationship was confirmed by Pavlova [264] for magnesium aluminosilicate glasses.

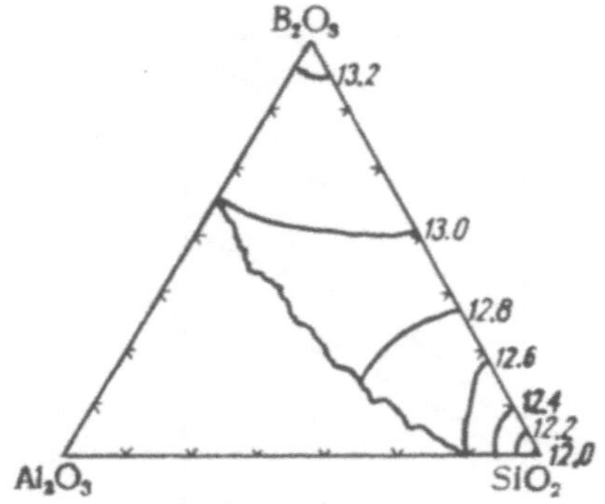

Fig. 45. Lines of equal log resistivity (isoresists) at 300°C for alkali-free glasses containing 30 mole % BaO and 70 mole % ($Al_2O_3 + B_2O_3 + SiO_2$). The diagram is a modification of an earlier publication [86], based on more precise data.

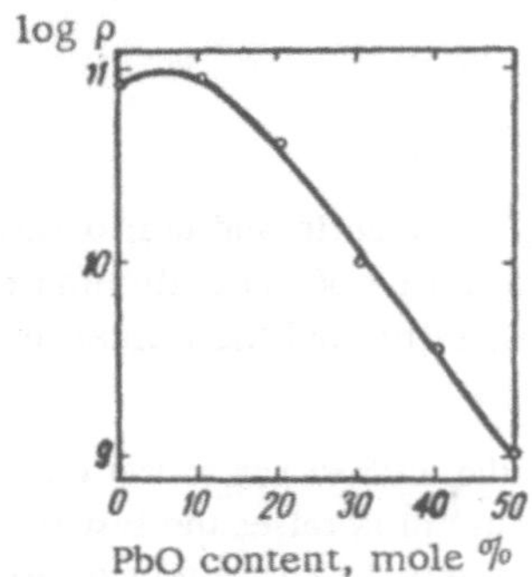

Fig. 46. Effect of replacement of BaO by lead oxide on the resistivity at 300°C of glass containing 50 mole % RO and 50 mole % SiO_2. Data of Mazurin and E. K. Mazurina.

Replacement of boric anhydride by alumina in alkali-free borate glasses leads, in distinction from silicate glasses, to little change of resistivity. It is interesting to note that the effect of replacement of B_2O_3 by Al_2O_3 differs at different temperatures (Fig. 43).

Replacement of silica by boric anhydride (Fig. 44) also raises the resistivity, with the effect strongest for the first few percent. This follows, in particular, from [150]. According to Mazurin and Lev [89] and Mazurin and Zubkova, partial replacement of boric anhydride by alumina has little effect on resistivity, lowering it somewhat in some cases and raising it somewhat in others.

Finally, the resistivities of glasses containing all three glass-forming oxides simultaneously were investigated (Fig. 45). It is seen that partial replacement (the first 10-15%) of SiO_2 by either of the other two glass-forming oxides produces a sharp increase of resistivity, with the effect of alumina being stronger than that of boric anhydride. In glasses containing not less than 15% of sesquioxides any mutual replacement of glass-forming oxides has very little influence on resistivity.

Influence of Simultaneous Presence of Bivalent Oxides. This was first investigated by Kuznetsov [59] for two series of glasses, and by Strauss et al. [226] for a series of lead glasses. It was subsequently studied with more extensive material by Mazurin and co-workers. When one bivalent oxide is partially replaced by another, the resistivity of the resultant glass is either higher than or corresponds to the value calculated from the additivity rule. Deviations from additivity increase with the difference between the sizes of the bivalent ions in the glass. The observed effects are usually much smaller than the polyalkali effect in high-alkali glasses. An example of the effect is shown in Fig. 46.

The resistivity of lead glass is very much increased if PbO is partially replaced by beryllium oxide. However, as no data on two-component beryllium silicate glasses are available, it is difficult to establish how much of this effect is due to simple approach to the composition of a pure beryllium glass, and how much to partial substitution.

Effects of Various Oxides. Figure 47 shows the effects of replacement of PbO by oxides of iron, cobalt, and nickel on the resistivity of lead glasses. Whereas nickel oxide has very little effect (the increase of resistivity in this case is obviously due merely to a decrease of the PbO content), the resistivity is lowered sharply by introduction of cobalt oxide and especially of ferric oxide. Ferric oxide has similar effects in lead borate and lead phosphate glasses [40] as well as in calcium aluminosilicate glasses [81].* Cobalt oxide does not lower the resistivity of glasses which do not contain lead.

According to our results, manganese oxide also lowers the resistivity of alkali-free glasses, but much less so than ferric oxide.

Milnes and Isard [273] have shown that the conductivity of lead silicate glasses increases with increasing water contents.

*See also [252, 258] with reference to the influence of ferric oxide on the conductivity of alkali-free glasses.

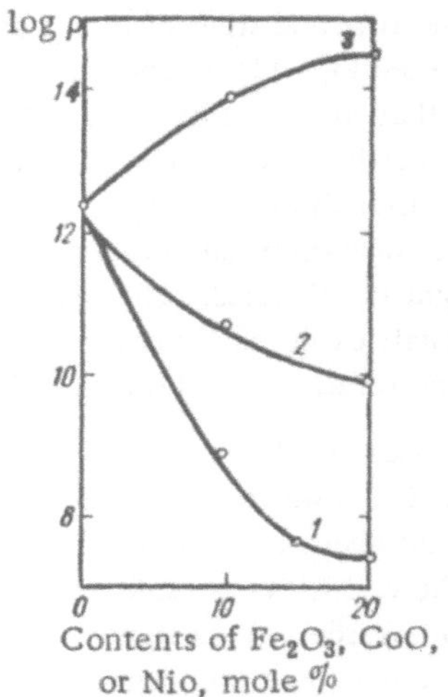

Fig. 47. Effect of replacement of lead oxide by various oxides on the resistivity at 150°C of glass corresponding to lead metasilicate in composition (data for nickel glasses are extrapolated from higher temperatures). 1) Fe_2O_3; 2) CoO; 3) NiO. Data of Mazurin and V. B. Brailovskii.

General Conclusion for Alkali-Free Glasses. It is evident from a comparison of data for alkali-free and alkali glasses that, on the whole, the effect of composition on resistivity is very much greater in high-alkali than in alkali-free glasses. Variations of composition have more effect on the resistivity of alkali-free glasses which have relatively low absolute resistivities (of the order of 10^8-10^{12} ohm-cm at 300°C). Variations of the composition of glasses already having resistivities of 10^{13}-10^{14} ohm-cm at 300°C generally have negligible effects. It appears that we are approaching a certain limit of attainable resistivities, represented by the expression $\log \rho_t = -0.5 + 9000/(t+273)$.

Influence of Alkali Oxides on Resistivity of Alkali-Free Glasses. We have shown that variations of composition influence the resistivity of high-alkali and alkali-free glasses in various and sometimes directly opposite ways. In order to unify data on these two composition regions into a single system, we must consider intermediate compositions – low-alkali glasses. The most convenient way of doing this is by studying the resistivities of various alkali-free glasses when alkali oxides are introduced into them.

As was shown a long time ago by Kurtts [61] and confirmed many years later by Strauss et al. [226], replacement of lead oxide in lead silicate glasses by alkali-metal oxides has little effect on resistivity, and in some cases even increases it. This was regarded as a curious but little-understood characteristic of lead silicate glasses, associated with unknown peculiarities of the lead ion.

Mazurin and Lev [89] investigated the effects of addition of various alkali oxides to a variety of alkali-free glasses. Some 15 different series of glasses were studied. Essentially the same relationships were found for all the series. Some characteristic curves from this study are shown in Fig. 48. For comparison, the same diagram shows variations of the resistivity of silica glass with introduction of alkali oxide.

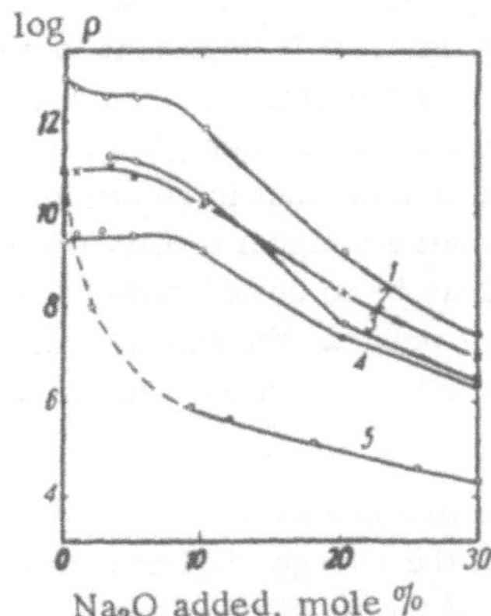

Fig. 48. Effect of addition of sodium oxide on the resistivity of alkali-free glasses at 300°C [89]. Original compositions (mole %): 1) BaO, 40; B_2O_3, 60; 2) PbO, 40; B_2O_3, 60; 3) BaO, 40; SiO_2, 60; 4) PbO, 40; SiO_2, 60; 5) SiO_2, 100.

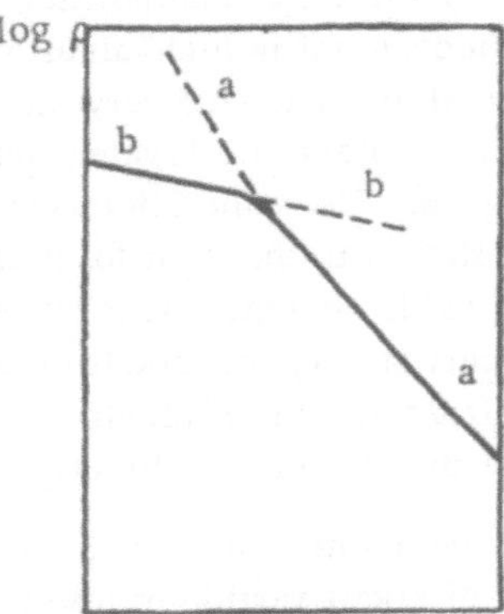

Fig. 49. General effect of added alkali oxides on the resistivity of alkali-free glasses containing oxides of heavy bivalent metals. a) Alkali conduction; b) nonalkaline conduction.

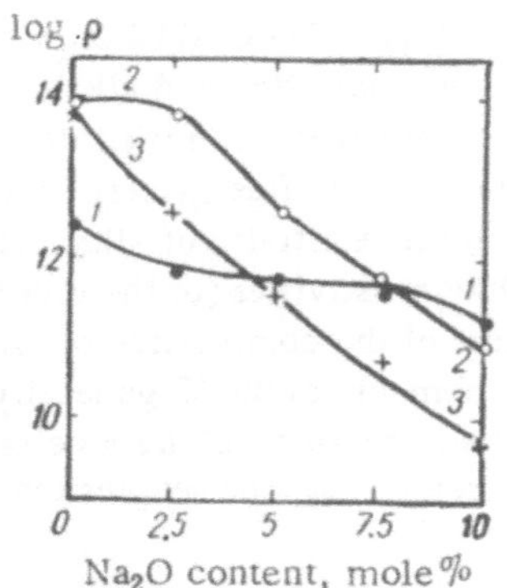

Fig. 50. Effect of replacement of silica by sodium oxide on resistivity in certain alkali-free glasses at 300°C. Original compositions (mole %): 1) BaO, 40; B_2O_3, 20; SiO_2, 40; 2) BaO, 30; Al_2O_3, 10; B_2O_3, 20; SiO_2, 40; 3) BaO, 30; Al_2O_3, 10; SiO_2, 60. Data of Mazurin, N. M. Zubkova, and V. A. Khar'yuzov.

It is seen in the figure that whereas the first additions of alkali oxide to silica glass (and also to boric anhydride – see Fig. 14) produce the greatest fall of resistivity, the first additions of alkali to alkali-free glasses containing considerable amounts of bivalent oxides have relatively little effect on resistivity. In the case of lead silicate glasses the resistivity even increases slightly at first, while the resistivities of barium and calcium glasses decrease somewhat; but the changes are slight in all cases. The resistivity first begins to fall rapidly with increasing alkali oxide content starting from 5-8% R_2O. The generalized form of the relationship is shown in Fig. 49.

The results obtained for low-alkali glasses are the natural consequence of the relations examined earlier. It is known from investigations of high-alkali glasses that introduction of a number of bivalent oxides such as CaO, BaO, PbO into these glasses raises the resistivity considerably; i.e., the mobility of the alkali ions is very much retarded. It is significant that the retarding effect of bivalent ions becomes more pronounced with lower contents of alkali-metal ions in the original glass. On the other hand, we know that increase of the bivalent oxide content in an alkali-free glass raises the conductivity, which is evidently not the consequence of movement of alkali-metal ions and which we describe as nonalkaline. It is therefore highly probable that if we have a glass of low alkali oxide content and introduce progressively increasing amounts of bivalent-metal oxides into it (predominantly with large ions) we will eventually obtain a glass in which alkali conduction is lower than nonalkaline. A change of the alkali ion content of such a glass would no longer alter the amount of principal charge carriers in the glass. In this case introduction of alkali oxides would, of course, have some kind of influence on the mobility of the bivalent ions or electrons causing conduction in the glass, but this influence would probably be relatively weak.

Accordingly, the curve in Fig. 49 can be divided into three regions: region of nonalkaline conduction, region of alkali conduction, and an intermediate region in which both types of conduction play an important part.

If we examine the conduction of low-alkali glasses in the light of these views, we can easily interpret the somewhat complex results obtained in this region.

The main effects of composition characteristic of alkali-free glasses should be observed to the left of the break; the effects usual in high-alkali glasses should be found to the right. It is important to note that the position of the break must depend very much on the composition of the original glass. It should shift to the left with increasing content of a bivalent oxide which raises the resistivity of alkali-free glass and lowers that of high-alkali glass. The same effect should occur with increasing alumina content in an original silicate glass. The break is shifted to the right for potash glasses, which have considerably higher resistivities than soda glasses at low alkali oxide contents. It is evident that the shift to the left may be so considerable in some cases that the break occurs at very low alkali oxide contents, sometimes even lower than the impurity content of the glass; on the very first additions of alkali to such glass we would observe alkali conduction, and the resistivity would then decrease sharply starting directly at the ordinate.

It must be pointed out once again that the retention of nonalkaline conduction in a glass in no way means that addition of alkali would not lower the resistivity. In some cases the change of glass composition due to the introduction of an alkali oxide may result in a definite increase of the mobility of the particles responsible for transfer of electricity in alkali-free glass. However, this decrease of resistivity must in general be very much less than the decrease caused by a corresponding increase of alkali ion content in a glass with alkali conduction.

The foregoing views are illustrated by two graphs, obtained by us jointly with N. M. Zubkova and V. A. Khar'yuzov. The first (Fig. 50) represents the different effects of an alkali oxide on the resistivities of alkali-free glasses of different compositions. It is seen that in borosilicate glass the transition from one type of conduction to the other occurs at 8-9% Na_2O. When 10% of BaO in this glass is replaced by alumina, the break is displaced far toward lower alkali oxide contents. In a pure silicate glass containing alumina the transition from one type of conduction to the other shifts still further to the left and occurs at about 0% Na_2O. Therefore, this series of glasses exhibits alkali conduction after the very first additions of alkali oxide.

The same results are shown in Fig. 51 in the form of curves representing the effect of gradual replacement of silica by barium oxide at constant alkali oxide content. At low BaO contents the glasses exhibit conduction of the alkali type. With increase of the bivalent oxide content the alkali conduction decreases and becomes first comparable to and then lower than the nonalkaline conduction. It is known that nonalkaline conduction increases with increasing BaO content. Therefore the resistivity is highest in the region of transition from alkali to nonalkaline conduction. It is clearly seen that the maxima shift toward lower BaO contents with decrease of the alkali oxide content. The maxima for potash glasses are far to the left of the maxima for soda glasses. This is because potash glasses have much higher resistivity in the region of alkali conduction, but the two types have approximately equal resistivities in the region of nonalkaline conduction.

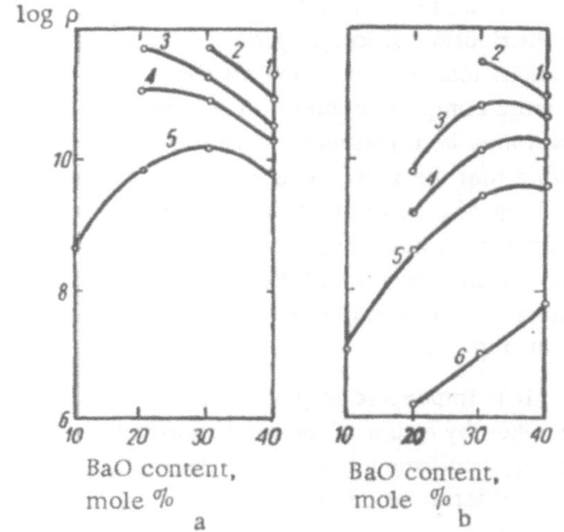

Fig. 51. Effects on resistivity at 300°C of replacement of silica by barium oxide in three-component potassium–barium silicate (a) and sodium–barium silicate (b) glasses. Alkali oxide contents (mole %): 1) 0; 2) 2.5; 3) 5; 4) 7.5; 5) 10; 6) 20. Data of Mazurin, N. M. Zubkova, and V. A. Khar'yuzov.*

Nature of the Influence of Composition on Conductivity of Glasses with Alkali Conduction

One possible interpretation of the influence of composition of high-alkali glasses on conductivity is given below. Because of the inadequacy of the experimental data, attempts to interpret the influence of composition on conductivity of glasses of other compositions are premature.

Other interpretations of similar material may be found in appropriate publications [58, 102, 105, 108-115, 117, 196-198]. Our interpretation is less detailed but covers more extensive experimental data.

Let us first consider certain aspects of the structure of glass [139, 141].

Most investigators consider that glass is a heterogeneous substance. A carefully made glass can contain two types of microheterogeneities: physical, caused by different degrees of order in the glass lattice in different regions, and chemical, resulting from nonuniform distribution of various components in the glass.

Complete crystallization (maximum order) of alkali glasses results in a very great increase of resistance. Therefore the ordered regions appearing in the glass take virtually no part in the conduction process. Consequently, it is permissible to disregard the possible existence in glass of regions with completely ordered structure (crystallites) when we examine the influence of composition on conductivity.

The problem of chemical heterogeneity of glass, formulated most clearly by Myuller [105], is much more important.

There is no doubt about the microheterogeneity of one region of the sodium borosilicate system [44, 141, 275, 276]. However, direct or indirect evidence of chemical heterogeneity of ordinary silicate glasses is especially interesting [15, 16, 39, 44, 133, 144, 147, 244, 248].

The nature of the heterogeneous structure of two-component alkali glasses is quite clear: the glass contains polar regions with alkali oxide concentration above the average, and nonpolar regions where the alkali

*Analogous relationships hold for the system $Na_2O-PbO-SiO_2$ [249].

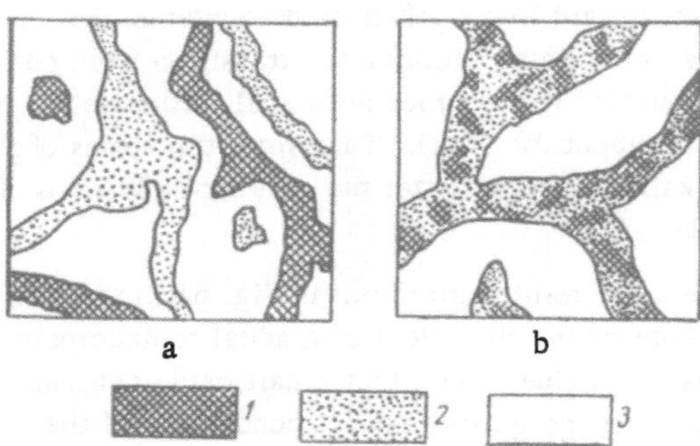

Fig. 52. Schematic representations of possible structures of glasses containing two different polar components. 1) Polar regions formed by the first polar component; 2) polar regions formed by the second polar component; 3) nonpolar regions.

oxide concentration is below average or even zero. The distribution of ions in glasses containing different alkali ions, or alkali and alkaline-earth ions, is a more complex problem. The structure of such glasses may be represented in various ways. It is possible that the glass contains polar formations, each consisting of dipoles of one kind only, separated by a nonpolar medium (Fig. 52a). It is also possible, however, that the primary differentiation is separation of the glass into a single polar and a nonpolar component (Fig. 52b).

It is impossible at present to perform calculations whereby either of these schemes could be conclusively confirmed or refuted. Only purely qualitative considerations can be applied.

We consider that variant b is the more probable. It is quite obvious that if a nonpolar medium contains two dipoles differing in size or charge, attraction and not repulsion forces must arise between them. In the same way, two associated groups, each formed from dipoles of one kind, different from the dipoles in the other group, must attract each other. Because of the mutual polarizing action, the forces of repulsion between them must be less than the forces of attraction.

The existence of an enormous number of ternary crystalline compounds containing neighboring actions differing in size and sign indicates that dipoles formed by two different binary silicates do not necessarily repel one another. It is definitely incorrect to say that the structure of a glass must necessarily be close to the structures of the primary crystalline phases. However, the possibility of energetically advantageous disposition of cations differing in size and charge in immediate proximity to each other must also be taken into account.

We now turn to a direct evaluation of experimental data.

The increase in the conductivity of alkali silicate and alkali borate glasses with increasing alkali oxide content is quite natural. The influence of alkali oxides is especially clear in silicate glasses. Increase of the R_2O content results in an increase of the amount of electricity carriers in the glass, which must lead to an increase of the statistical factor, as represented by the expression

$$A = \frac{n_0 q^2 \delta^2 \nu}{3kT} \tag{16}$$

The effective activation energy must decrease at the same time.

Data on the sodium silicate system, which has been studied in most detail, are presented in Table 5.

It must be taken into account that the usual scatter of the experimental values of A is very great; it may be 100% or more.

Evidently the bond strength is least between potassium and oxygen ions, and greatest between lithium and oxygen ions. Consequently, the lithium ion should have the highest dissociation energy. At the same time, the activation energy should be highest for the large potassium ion, and lowest for the lithium ion. It is evident from Table 6 that in the transition from potassium silicate to lithium silicate glasses the change of activation energy plays a more important part than the change of dissociation energy.

There is reason to believe that introduction of alkali ions into borate glasses changes the boron–oxygen framework: boron passes from threefold to fourfold coordination [158, 267, 172]. In boric anhydride each boron ion is surrounded by three oxygen ions. When alkali-metal oxides are introduced into boric anhydride, additional oxygen ions appear in the glass. Some of the boron ions become surrounded by four oxygen ions, so that the center of the boron–oxygen tetrahedron acquires a single negative charge. The alkali-metal ion is therefore bonded electrostatically with the center of the boron–oxygen tetrahedron [108].

TABLE 5. Effects of Composition on the Effective Activation Energy and Statistical Factor of Glasses in the System Na_2O-SiO_2 *

Glass composition, mole %		U_0, eV	log A
Na_2O	SiO_2		
5	95	0.98	1.26
13	87	0.70	1.33
20	80	0.68	1.54
25	75	0.63	1.66
30	70	0.59	1.54
36	64	0.53	1.49
40	60	0.51	1.57
45	55	0.52	1.93
48	52	0.47	1.68

*Based on the data of Mazurin [73] and of N. G. Suikovskaya, K. K. Evstrop'ev, and Mazurin.

TABLE 6. Effect of the Kind of Alkali-Metal Ion on the Activation Energy of Certain Simple Silicate Glasses *

Glass composition, mole %	U_0, eV
Li_2O—33.3; SiO_2—66.6	0.61
Na_2O—33.3; SiO_2—66.6	0.55
K_2O—33.3; SiO_2—66.6	0.64
Li_2O—20.0; SiO_2—80.0	0.68
Na_2O—20.0; SiO_2—80.0	0.68
K_2O—20.0; SiO_2—80.0	0.73

*From the data of Mazurin et al.

TABLE 7. Effect of the Composition of Fluoride Glasses on Resistivity and Activation Energy [128]

Glass composition, mole %						log ρ		U_0, eV
BeF	AlF_3	CsF	CaF_2	SrF_2	BaF_2	150° C	250° C	
70	10	20	—	—	—	12.92	8.35	2.00
60	10	30	—	—	—	11.23	6.90	1.90
50	10	40	—	—	—	10.40	6.84	1.56
60	10	20	10	—	—	15.89	10.81	2.22
60	10	20	—	10	—	14.36	9.74	2.02
60	10	20	—	—	10	13.83	9.20	2.02

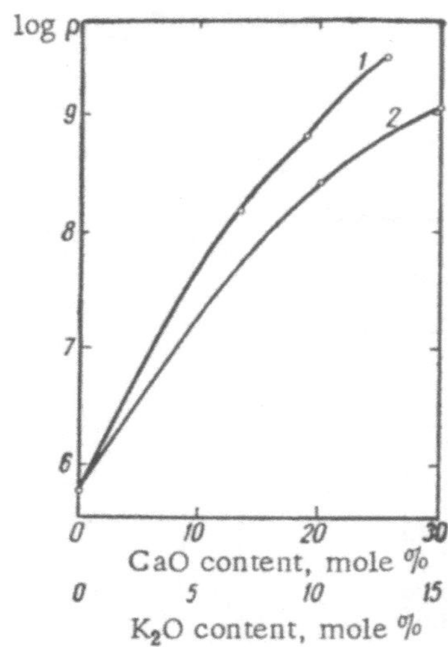

Fig. 53. Effect of replacement of silica on the resistivity at 150°C of glass containing 20 mole % Li_2O and 80 mole % SiO_2. 1) Replacement by potassium oxide; 2) replacement by calcium oxide. Data of Mazurin and M. I. Kalinin [76].

This greatly complicates all the conductivity relations, especially in the low-alkali region. The conductivity of two-component alkali borate glasses has been subjected to most detailed theoretical analysis by Myuller [105, 115], and is not considered here.

In a discussion of the nature of the polyalkali effect, we must first state the basic principle on which this effect is interpreted, formulated as long ago as 1953 [73]. In a glass containing two different alkali ions of different sizes each ion moves predominantly along the vacancies left by ions of the same size. Therefore, for example, introduction of lithium ions into a soda glass cannot substantially facilitate the movement of sodium ions by increasing the number of possible vacancies. The total amount of electricity carriers is determined in practice by the amount of the alkali oxide present in the greatest concentration in the glass. This principle is indirectly confirmed by studies of the nature of conduction in glass (see p. 6), not only in the case of replacement of a smaller by a larger ion but also, most significantly, in the case of replacement of a larger by a smaller ion. For example, as has been emphasized by a number of investigators, it is much more difficult to replace electrolytically sodium by lithium than sodium by sodium in soda glass.

It follows from the principle of independent movement of alkali ions that the conductivity must decrease when one alkali oxide is replaced by another. If we assume that no interaction occurs between alkali-metal ions of different sizes, the conductivity $\varkappa$ of a glass containing, for example, m% R'_2O, n% R''_2O, and (100-m-n)% SiO_2 is given by the expression

$$\varkappa = \varkappa' + \varkappa''$$

where $\varkappa'$ is the conductivity of glass containing m % R'_2O (rest SiO_2) and $\varkappa''$ is the conductivity of glass containing n % R''_2O (rest SiO_2).

If the conductivities $\varkappa'$ and $\varkappa''$ were linear functions of the R'_2O and R''_2O contents, the variation of $\varkappa$ with the ratio m/(m+n) at constant (m+n) would also be linear. In reality, the functions $\log \varkappa' = f(m)$ and $\log \varkappa'' = f(n)$ are approximately linear (see Figs. 13 and 14); the inevitable result is that the conductivity of complex glasses exhibits a minimum.

However, it is found experimentally [76] that the conductivity decrease caused by partial replacement of one alkali ion by another is very much greater than is to be expected from the principle of independent movement of alkali ions. Therefore, for a more precise description of the effect we must take into account the influence of the less mobile ions on the mobility of the more mobile ones. It is logical to assume that all modifier ions, whether they are uni-, bi-, or trivalent, should on the whole have the same effect on the mobility of the electricity carriers (i.e., the alkali ions present in the highest concentration) in the glass.

In particular, if small amounts of alkali-metal ions of a particular size are introduced into a glass containing alkali-metal ions of a different size, the former take virtually no part in the transport of current. They merely influence the mobility of the charge carriers.

We must therefore consider the influence of modifying ions on the mobility of alkali ions in silicate glass. In view of the peculiarities noted earlier, the relations for borate glasses must be more complex.

Figure 53 shows the effect of replacement of SiO_2 by potassium and calcium oxides on the conductivity of lithium silicate glass. With equal replacement of SiO_2 by modifying oxides, the amount of alkali ions introduced is double the amount of bivalent ions. The abscissa scale is therefore chosen so that the amounts of modifying ions introduced are equal in both cases.

It is clear from Fig. 53 that the curves for uni- and bivalent ions are quite similar.

Let us consider the probable physical interpretation, which we proposed jointly with G. T. Petrovskii, of the influence of modifying ions on conductivity [78, 84, 85]. (Similar views have been advanced by K. S. Evstrop'ev.)

Oxygen ions occupy a considerable proportion of the volume of any silicate glass. They surround silicon ions, and alkali and alkaline-earth ions. Alkali ions in silicate glass are contained and move in an "oxygen atmosphere," and interact with other positive ions through oxygen ions rather than directly. Therefore each alkali ion is bonded to all its oxygen environment and the strength of its bonding in the cavity which it occupies is determined by its ability to interact with the surrounding oxygen ions. The bond strength with each oxygen must depend primarily on the state of that oxygen. For example, if an oxygen ion is bonded to two silicon ions its electron shell is so strongly polarized by the silicon ions that its bonding to the alkali ion is negligibly weak. If the oxygen ion is linked by one valence to a uni- or bivalent ion, the situation is radically altered. The weakly polarized oxygen ion may be additionally polarized by the alkali ion under consideration and thereby increase the strength of the bond of the alkali ion with the position where it is held. The larger the size or the smaller the charge of the neighboring modifying ion, the weaker is the polarization of the oxygen ion and the more firmly is the alkali ion held near this oxygen. We thus have a logical explanation of the influence on the conductivity of the size of the bivalent ion introduced into glass (see pp. 25-26).

In our explanation of the influence of extraneous modifying ions on the mobility of alkali ions it was assumed that both these kinds of ions have the same oxygen ions as neighbors; this follows from the view put forward earlier that all polar groupings in the glass are combined within single polar formations. Of course, definite additional differentiation of different polar groups is possible within these groupings. However, in the case of movement over long distances, characteristic of through conduction, an alkali ion must travel past all the polar ions present in the glass. These concepts can be confirmed by studies of glasses in which the charge carriers are bonded to uni- and bivalent ions directly rather than through oxygen.

As was shown in [234], crystalline fluorides exhibit anionic conduction. Therefore fluoride glasses probably constitute a rare class of systems with anionic conduction. Table 7 contains certain data on the influence of composition on the conductivity of fluoride glasses [128]. It follows from [128] that the effect of alkali fluorides on the conductivity of alkali-free fluoride glasses is entirely analogous to the effect of fluorides of bivalent metals; this is in good agreement with the concept of anionic conduction. In most cases the size of the ion has a distinct influence; this applies equally to both uni- and bivalent ions. However, this influence is directly opposite to the corresponding influence in silicate glasses. The larger the ion, the less its ability to polarize neighboring fluoride ions, the higher is the conductivity of the glass. Fluoride glasses do not contain an "intermediary"; the bonding forces between the positive and negative ions directly determine the mobility of these negative ions.

It is noteworthy that, according to [128], the resistivity of fluoride glasses is raised somewhat by partial replacement of one alkali oxide by another. Alkali oxides do not take part in transfer of electricity in this case, and therefore the polyalkali effect cannot occur. Apparently the resistivity is increased owing to a factor which could be described as the multicomponent effect. When new ions, differing in size from those already present, are introduced into a glass, conditions arise for closer "packing" of the ions in the glass. This factor may lead to an increase of the effective activation energy (the charge carriers meet with more difficulty in "breaking through" the lattice) and therefore of the resistivity. The multicomponent effect may, in particular, explain why the resistivity of alkali glasses is increased above the additive value by partial replacement of one bivalent oxide by another (see Table 3).

In the light of the foregoing, the polyalkali effect is merely a consequence of the general relations found for the influence of extraneous modifying ions on the mobility of alkali-ion charge carriers in glass. Indeed, the polyalkali effect should depend both on a decrease of the content of the main charge carrier and on the influence of the extraneous modifying ion introduced into the glass, which works to retard the movement of the main charge carriers.

The principle of independent movement of alkali ions can be fully applicable only to pairs of ions of very different sizes. The closer the radii of the bivalent ions, the less pronounced must the polyalkali effect be.

TABLE 8. Effect of Partial Replacement of One Alkali Oxide by Another on log A [73]

Glass composition, mole %				log A
SiO_2	Li_2O	Na_2O	K_2O	
66.6	33.3	—	—	1.94
66.6	—	33.3	—	1.51
66.6	—	—	33.3	1.91
66.6	16.7	16.7	—	3.51
66.6	16.7	—	16.7	2.41
66.6	—	16.7	16.7	3.82

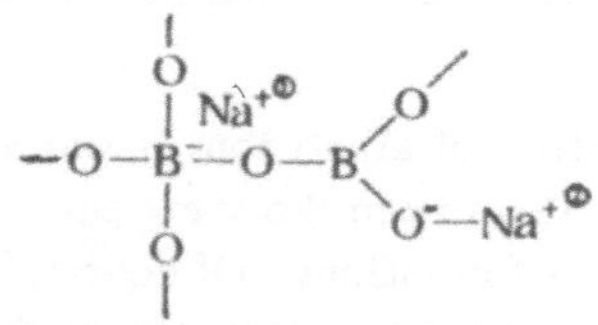

Fig. 54. Alternative positions on the Na^+ ion in the borate glass network.

In sodium–potassium and lithium–sodium systems movement of ions of one size along the vacancies left by ions of the other size is also apparently possible to a limited extent. This was first pointed out by Myuller [105]. This possibility is virtually excluded for lithium–potassium and even more so for lithium–cesium systems; this accounts for the difference between the absolute values of the effects.

One more relationship should be noted. If the amount of electricity carriers is kept constant, introduction into the glass of any ions which do not carry electricity but which increase the activation energy leads to an increase of the statistical factor, while introduction of ions (such as aluminum) which decrease U_0 (but take no part in transport of electricity) decreases the statistical factor. It is possible that changes of the distances between the network points with changes of glass composition play a certain part in variations of A.

It should also be taken into account that a change of composition must also alter the vibration frequency ν of the ion, which is contained in the numerator of the expression (16). With increased strength of bonding of the ion in the equilibrium position (i.e., with increase of U_0) ν should increase, leading to increase of A.

In the great majority of cases A is altered by a factor of 6-8 at most by introduction of extraneous ions. We consider that it is quite possible to attribute this alteration to the two factors discussed above.

Abnormally large changes of A are found for lithium–sodium and sodium–potassium glasses (Table 8). It is possible that entropy changes during dissociation should be taken into account in order to explain such large increases of A; this was proposed by Myuller [117], and we refer those interested in the matter to his work. It should be noted that the increase of A for lithium–potassium and especially lithium–cesium systems is very much less than for lithium–sodium and sodium–potassium systems. A similar relation holds for borate glasses [6, 98].

It is interesting to note that while the variations of conductivity of alkali borate and silicate glasses with increase of alkali oxide content are entirely different from each other (see Figs. 13 and 14) the curves representing conductivity variations of silicate and borate glasses due to the polyalkali effect are very similar (see Fig. 19).

Equimolecular replacement of one alkali oxide by another in borate glass maintains a strictly constant ratio between nonpolar and polar boron structural units, and therefore the structure of the glass-forming oxide network is retained. We should then observe without any interfering effects the results of mutual action of different alkali-metal oxides; evidently, this should not differ in any significant way from the results of interaction of the same ions in a silicate glass. Increase of the R_2O content, on the other hand, affects the structures of silicate and borate glasses differently, and this influences the respective conductivity relationships.

Glasses of R_2O–RO–B_2O_3 systems have been studied little. The processes which take place in them during changes of composition are very complex.

TABLE 9. Comparison of Activation Energies U_0 for Sodium Silicate and Borate Glasses

Glass composition, mole %			Na_2O concentration, M	U_0, eV
Na_2O	SiO_2	B_2O_3		
10	90	—	3.78	~0.75
10	—	90	2.98	1.36
20	80	—	7.93	0.67
20	—	80	6.44	0.95
30	70	—	12.20	0.59
30	—	70	10.40	0.75

In contrast to silicate glasses, in borate glasses the sodium ion can occupy two alternative positions: near the center of a boron–oxygen tetrahedron or near its periphery (Fig. 54). It may be assumed that the alkali ions at the periphery of the tetrahedrons are bonded less firmly than those at the centers of the boron–oxygen tetrahedrons; this accounts, in particular, for the higher conductivity of silicate glasses as compared with analogous borate glasses. It may also be assumed that bivalent ions introduced into an alkali borate glass displace some of the alkali ions from their positions at the centers of the boron–oxygen tetrahedrons and occupy these positions themselves. It follows that introduction of bivalent ions must increase the proportion of alkali ions at the peripheries of the tetrahedrons, and this should lead to increased conductivity. The greatest effect should be produced by small bivalent ions, which are held most strongly by the negatively charged centers of the tetrahedrons. This effect does not appear at low alkali contents – the probability of formation of new tetrahedrons is not high enough. At very high R_2O contents many alkali ions are already at the peripheries and some increase in their number does not have any significant influence on conductivity. With moderate R_2O contents the first portions of the introduced bivalent oxide may produce a very large relative increase of the amount of these "peripheral" alkali ions in the glass. The effect becomes weaker as more RO is introduced, and the usual effect of the bivalent ion on conductivity of alkali glasses (see Fig. 26) becomes predominant.

In conclusion, let us examine the effect of partial replacement of SiO_2 by glass-forming oxides on the conductivity of silicate glasses. Experimental data show that such replacement of SiO_2 has much less effect on resistivity than replacement of SiO_2 by modifying oxides. Apparently, if the same ion can act both as a glass-former and as a modifier, it should allow more freedom of movement to the alkali ions in the former case than in the latter. This may be attributed to changes in the coordination number of oxygen characteristic for the ion in the two positions. When acting as a glass-former, the positive ion is bonded with fewer oxygen ions (usually four) and therefore has a much stronger polarizing effect on them; as already noted, this should lead to weakening of the bonds between the alkali ions and these oxygen ions. If the positive ion is in sixfold coordination, its polarizing effect on each oxygen ion is much weaker, and the interaction between the oxygen and sodium ions is intensified. It is also possible that the packing density of the glass network, which becomes higher with increase of the coordination numbers of the ions present in the glass, plays a definite part. The effects of the oxides of Be, Mg, and Zn on the conductivity of alkali glasses (see Fig. 22) may be considered in this context. When bivalent ions of small size are introduced into a high-alkali glass they can apparently act, at least partially, as glass-formers, and the effect of replacement of SiO_2 by these oxides is slight. When the amount of such oxides added exceeds a certain percentage, which depends primarily on the amount and nature of the alkali oxides which provide the bivalent ions with the additional oxygens necessary for formation of the tetrahedral framework, the corresponding ions begin to act exclusively as modifiers in the glass. As a result, the curve should show a break with a subsequent greater increase of resistivity. It is evident that the lower the alkali oxide content the lower the RO content at which this break appears (see Fig. 23). The lower the polarizing power (and therefore the ability to interact with oxygen) of the alkali ion in the glass, the more pronounced should be the glass-forming action of bivalent oxides. Therefore the effect due to the glass-forming properties of Mg and Zn is most prominent in potassium glasses and does not appear at all in lithium glasses. At the same time, Be can act as a glass-former even in lithium glasses.

The weak effect of titanium dioxide on glass resistivity can probably be explained on the same grounds.

Curves representing the influence of aluminum oxide can, in general, be interpreted similarly. As long as the alumina introduced into the glass enters the glass network (up to the ratio $Al_2O_3 : R_2O = 1 : 1$) it generally lowers the resistivity. Some writers [28, 200] attribute this effect to the larger size of aluminum—oxygen tetrahedrons in comparison with silicon—oxygen, allowing alkali ions to move more freely. When the amount of Al_2O_3 is increased further, the alumina probably enters the glass as a modifier and the resistivity begins to increase considerably. The initial increase of resistivity on addition of alumina can be attributed to the multicomponent effect. The magnitude of the initial maximum should diminish as the composition of the glass becomes more complex.

The glass-forming properties of lead ions are the consequence of the easy deformability of the electron shell of lead itself. Lead is not capable of strongly polarizing the electron shell of oxygen; yet the mobility of alkali ions is influenced only by the degree of deformation of the oxygen electron shell. Therefore the curves for lead glass are not of the form described above.

Let us now examine the curves representing the effect of replacement of SiO_2 by boric anhydride (see Fig. 29).

It is known that a borosilicate glass is separated into borate and silicate components. Alkali oxides tend to pass mainly into the borate component [26, 44]. This is in good agreement with conductivity data. Indeed, the effective activation energy, which characterizes the strength of the bonds holding the alkali ion in the glass network, is much lower for silicate glasses than for the corresponding borate glasses (Table 9). Accordingly, with equilibrium distribution of alkali oxides between the two components the amount present in the borate component should be considerably greater.

Thus, initial replacement of SiO_2 by boric anhydride "draws off" alkali ions strongly from the silicate component, which therefore becomes poorer in alkali. The resistance of the silicate component, which as yet determines the total resistance of the glass, increases substantially. When the amount of borate component becomes sufficient to form continuous bridges between the electrodes, its conductivity begins to play an increasingly important part in the total conductivity. Increase of the amount of B_2O_3 is not accompanied by an equivalent decrease of Na_2O content because with increase of the B_2O_3 content the borate component withdraws fresh amounts of alkali oxide from the shrinking silicate component. Consequently, the resistance of the borate regions shows little increase. The alkali concentration in the borate component is now much higher than in the silicate component and the conductivity of the glass is determined by the conductivity of the borate regions. As a result of all this the increase of resistivity slows down during further replacement. Finally, when nearly all the alkali-metal oxides are in the borate component, further increase of the B_2O_3 content of the glass leads to a proportional decrease of the Na_2O concentration in the borate regions, so that the change of resistivity during subsequent replacement becomes greater again.

Finally, it should be noted that, as is evident from all the experimental material at our disposal, studies of plots of log ρ versus concentration (or of log $\varkappa$ versus concentration) often reveal very simple relations. In a number of cases the influence of changes of composition is found to be additive. It is easy to show that very much more complex relations are obtained if ρ or $\varkappa$ is taken as the ordinate. K. S. Evstrop'ev demonstrated that this effect is associated with additivity of the effective activation energy of the charge carriers. The statistical factor A alters relatively little with composition. Therefore it follows from the fundamental conductivity equation (7) that in the first approximation log $\varkappa$ is inversely proportional and log ρ is directly proportional to U_0.

Method of Calculating the Conductivity of Glass

In investigations of any property of glass the problem of devising a method for calculating that property inevitably arises. As already noted, simple and exact formulas can be derived for calculating some properties of glass, while others can hardly be calculated at all in the general form. This latter group includes all the electrical properties, apart from the dielectric constant.

We consider that for properties which exhibit nonadditive relations approximate methods should be devised in some cases, so that it would be possible to estimate the order of magnitude of a particular property of a glass of known composition but unknown properties.

TABLE 10. Comparison of Calculated and Experimental Change of $t_{\varkappa-100}$ with Change of Glass Composition

Glass composition		$\Delta t_{\varkappa-100}$		
Initial	Final	Exptl., Gehlhoff and Thomas [171]	Calc., Gehlhoff and Thomas [171]	Calc., author's formulas
$Na_2O \cdot BaO \cdot 6SiO_2$	$Na_2O \cdot CaO \cdot 6SiO_2$	−45°	+24.5°	−28°
$Na_2O \cdot MgO \cdot 6SiO_2$	$Na_2O \cdot CaO \cdot 6SiO_2$	+60°	+134°	+55°
$K_2O \cdot BaO \cdot 6SiO_2$	$K_2O \cdot CaO \cdot 6SiO_2$	−35°	+30°	−24°
$Na_2O \cdot BaO \cdot 6SiO_2$	$K_2O \cdot BaO \cdot 6SiO_2$	+74°	Cannot be calculated	+76°

Apart from Ambronn's formula [156], which can be used for accurate calculations of glass conductivities in the system $Na_2O-CaO-SiO_2$, the only methods used at the present time for calculating conductivities of solid glasses are those of Gehlhoff and Thomas [171] and of Fulda [167]. The tables of Gehlhoff and Thomas purport to cover a wide range of glass compositions and have been included in all reference works and most text books for 30 years. Their tables and those of Fulda are quite similar, because Fulda adhered strictly to the method of compilation proposed by Gehlhoff and Thomas and used glasses made by them. We shall therefore comment on the tables of Gehlhoff and Thomas.

These tables cannot be used for calculating absolute conductivities and are intended only for calculations of changes of $t_{\varkappa-100}$ (the temperature at which the resistivity of the glass is 10^8 ohm-cm) with changes of composition. Some of the results used by Gehlhoff and Thomas in compiling the tables, especially data on the effect of calcium oxide, gave rise to serious doubts a relatively long time ago [73]. However, discrepancies between the results of two investigations can often be attributed equally to errors in either. Therefore the erroneous nature of the Gehlhoff and Thomas tables could be proved only by the use of their own data. The point is that, apart from the main investigations on which their tables are based, Gehlhoff and Thomas studied conductivities of a small series of glasses of the general composition $R_2O \cdot RO \cdot 6SiO_2$ and reported numerical values of $t_{\varkappa-100}$ for them (which they did not do for other glasses). The compositions of the glasses in this supplementary series were very close to those of the series on which the tables were based. Therefore excellent agreement between the calculated and experimental values was to be expected. The usual errors (often very considerable) due to differences in the melting and measurement techniques should be entirely absent.

We performed the appropriate calculations. We calculated from the tables the changes of $t_{\varkappa-100}$ in the transition from one glass to another, and compared the results with the differences between the experimental values. The results are given in Table 10. The table also gives the results obtained in similar calculations with the aid of new formulas proposed below.

With regard to the part relating to the Gehlhoff and Thomas calculations, Table 10 needs no detailed comment. It is quite obvious that their method of calculation gives not only large errors in the magnitude of the change of $t_{\varkappa-100}$ but even errors in the sign of that change. We attribute this mainly to errors in the synthesis of glasses in the system $Na_2O-CaO-SiO_2$ in the series for which the experimental results were used for compilation of the tables.

In our opinion, our calculations not only demonstrate that the Gehlhoff and Thomas tables are completely useless for calculating glass conductivities, but should also warn investigators using other tables published by the same authors, because they measured various properties for glasses of the same meltings.

Therefore only Ambronn's data are available to investigators at present. These data are reliable, but the range of compositions covered is too narrow.

Before attempting to devise a more general method for calculating conductivity, we must establish the relative validity of any principles on which such calculations may be based.

It is quite obvious that, since composition has a very strong and highly complex influence on conductivity of glass, one cannot hope to develop methods for exact calculation of the conductivities of solid glasses of complex composition.

TABLE 11. Comparison of Values Calculated from the Author's Formula with Experimental Data for Certain Simple Glasses*

Glass comp., mole %				log ρ					
				150° C			300° C		
SiO_2	MgO	Na_2O	K_2O	Exptl.		Calc.	Exptl.		Calc.
				1st melting	2nd melting		1st melting	2nd melting	
70	13	17	—	7.23	7.28	7.07	4.91	4.99	4.87
70	13	15	2	8.09	8.25	8.29	5.43	5.60	5.68
70	13	13	4	8.76	8.98	9.20	5.86	6.14	6.28
70	13	11	6	9.41	9.52	9.86	6.42	6.43	6.72
70	13	8.5	8.5	10.29	10.11	10.29	7.07	6.87	7.02
70	13	3	14	9.67	—	9.68	6.68	—	6.68
70	13	—	17	8.09	7.97	8.55	5.72	5.65	5.97
73	—	27	—	5.87	—	5.76	3.94	—	3.83
73	—	20.25	6.75	7.93	—	7.97	5.14	—	5.17
73	—	13.5	13.5	9.06	—	8.81	5.83	—	5.70
73	—	6.75	20.25	8.29	—	8.32	5.43	—	5.44
73	—	—	27	6.57	—	6.50	4.28	—	4.38
	CaO								
87	—	13	—	6.97	—	7.19	4.81	—	4.96
82	5	13	—	7.97	—	7.75	5.51	—	5.41
72	15	13	—	9.18	—	8.96	6.36	—	6.31
67	20	13	—	9.58	—	9.59	6.66	—	6.76
62	25	13	—	10.31	—	10.19	7.26	—	7.21

*Data of T. M. Makarova, Mazurin, and V. S. Molchanov, and of [76].

However, users are mainly interested in the electrical properties of glasses in relation to the possible use of these glasses as insulators. Therefore in most cases it is important for practical purposes to determine merely the order of magnitude of the resistivity of a given glass at a given temperature. This explains, in particular, why specifications for vacuum tube glasses give values only of $t_{\varkappa-100}$, i.e., values which are only very rough indications of the resistivity at temperatures differing substantially from $t_{\varkappa-100}$ itself. Therefore, for properties such as conductivity, calculation methods which do not ensure a high degree of precision may also be useful. If the difference between the calculated and experimental values of log ρ does not exceed one order of magnitude, the precision of the calculation may be regarded as satisfactory. At the same time, the calculation method should cover a wide range of temperatures. The resistivity calculations should be more accurate at relatively high temperatures because at or near room temperatures the volume resistivity of most glasses used in practice is high and surface conductivity is the main factor in leakage of current in glass insulators. The method should provide single formulas covering a wide composition range. Finally, it should not involve excessively long and complicated calculations.

Calculation formulas may be adapted for glass compositions expressed in either weight or molecular percentages. In general, calculations based on weight percentages are more convenient for practical purposes. However, experiment shows that the relations between log ρ and composition are more complex if the latter is expressed in weight percentages. Therefore we devised a method for calculating log ρ from compositions expressed in molecular percentages.

A possible method of calculating the resistivity of solid glass is described below. The proposed method is purely empirical. No physical meaning should be ascribed to the numerical coefficients.

The method is subject to the following limitations with regard to composition (in molecular percentages): the sum of ($Na_2O + K_2O$) must be in the range of 12-30%, with the two oxides in any proportion; the method is not applicable to glasses containing lithium oxide; the total RO content (where R is Mg, Zn, Pb, Ca, Ba) must be in the range of 0-20%, with any proportions of the bivalent oxides. If the only bivalent oxides present are

TABLE 12. Verification of the Author's Formula for Certain Vacuum Tube Glasses

Glass code	$t_{\varkappa-100}$, °C	
	Exptl.	Calc.
ZS-4	332 [153]	327
№ 2	178 [153]	157
BD-1	224 [153]	237
№ 23	240 [153]	270
№ 16	158 [94]	167

BaO, CaO, or both, the RO content can be up to 25-28%. With some increase of error, the resistivities of glasses containing BeO (taken as MgO in the calculation) and SrO (taken as BaO) can be calculated. In addition, the Al_2O_3 and B_2O_3 contents can be in the range of 0-10%; up to 0.5% of other oxides (apart from Li_2O) can be disregarded; the rest of the glass should consist of silica.

Accuracy of the Method and Temperature Limits of Applicability. Our calculations (a few results are given in Tables 11 and 12) showed that between the softening temperature and 150°C the difference between calculated and experimental values of log resistivity does not exceed half an order of magnitude both for relatively simple and for complex industrial glasses. We therefore consider that in this temperature range accuracy of within an order of magnitude can be guaranteed provided that correct measurement techniques are chosen. This last condition is very important, as in some instances values found by different workers for the resistivity of the same glass differ by several orders of magnitude. The method becomes less accurate for calculations of resistivity at temperatures below 150°C, but even in the worst cases the error should not be much more than an order of magnitude down to 100°C. As already noted, at lower temperatures surface conductivity is the predominant factor in the insulating properties of glass.

Description of the Method. Formulas for calculating resistivity at 300°C are given. If resistivities at other temperatures are required, log A is also calculated. This is the intercept cut off along the ordinate at 1/T = 0 by the extrapolated linear plot of log ρ versus 1/T. The value of log ρ at any temperature, and also $t_{\varkappa-100}$, can then be calculated from special formulas.

Calculation of log Resistivity at 300°C (log ρ_{300})

Symbols:

a – total alkali oxide content;
a_K – K_2O content;
b – total RO content;
b_{MZ} – sum of MgO and ZnO contents;
b_C – CaO content;
b_{BP} – sum of BaO and PbO contents;
c – Al_2O_3 content;
d – B_2O_3 content.

The following formula is used for the calculation:

$$\log \rho_{300} = (75 - a) \times 0.08 + (38 - a) \times 0.05 \frac{a_K}{a} + \left[0.25 - \left(\frac{a_K}{a} - 0.5\right)^2\right] \times 6.4 + 0.018\, b_{MZ} + (30 - a)^2 \times \frac{b_C + b_{BP}}{7300} + 0.05\, b_C + 0.08\, b_{BP} - 0.05\, c + (30 - a)^2 \frac{d}{6000} + 0.04\, d + 0.015\, b^*$$

All the quantities in the formula are in molecular percentages.

*The last term in the formula is added only if the glass contains at least two different bivalent-metal oxides and the content of each exceeds 1%.

Calculation of log A

The following formula (the symbols have the same significance as before) is used:

$$\log A = (30 + a) \times 0.03 + \left[0.25 - \left(\frac{a_K}{a} - 0.5\right)^2\right] \times 0.22a + 0.03\, b_{BP} - 0.01c$$

Calculation of log ρ for Any Temperature (log ρ_t) and of $t_{\varkappa-100}$

$$\log \rho_t = (\log \rho_{300} + \log A) \times \frac{573}{t + 273} - \log A; \quad t_{\varkappa-100} = \frac{(\log \rho_{300} + \log A)\, 573}{8 + \log A} - 273$$

Calculation of Resistivity at Any Temperature from One Experimental Point

If the resistivity at one temperature (e.g., room temperature) is known from the literature or from one's own determinations, or (which happens very often) only $t_{\varkappa-100}$ is known, the resistivity at any temperature can be calculated with great accuracy by the proposed method. The basis of the calculation is that log A depends little on composition for the range of glasses under consideration. Moreover, it is quite obvious that errors in determination of log A would have relatively little effect on the calculated resistivities at moderate temperatures. Therefore in using the formulas given below for calculating resistivities from a single experimental point we can be certain that the accuracy of the calculation depends primarily on the accuracy with which the experimental value was determined.

Formulas are given below for calculating the resistivity at any temperature t and for calculating $t_{\varkappa-100}$ from an experimental value of resistivity at temperature t_e, and also for finding the resistivity at any temperature t from known $t_{\varkappa-100}$:

$$\log \rho_t = (\log \rho_{t_e} + \log A)\,\frac{t_e + 273}{t + 273} - \log A$$

$$t_{\varkappa-100} = \frac{\log \rho_{t_e} + \log A}{8 + \log A} \cdot (t_e + 273) - 273$$

$$\log \rho_t = (8 + \log A)\,\frac{t_{\varkappa-100} + 273}{t + 273} - \log A$$

The calculations are based on the following assumptions. In the first approximation, the effects of most bivalent and trivalent oxides are independent of the amounts present but depend very much on the contents of alkali-metal oxides; the polyalkali effect can, in the first approximation, be taken as independent of the contents of alkali oxides, although the increased difference between the resistivities of soda and potash glasses must be taken into account; the multicomponent effect is most pronounced in the presence of several bivalent-metal oxides; the influence of individual oxides and effects on the logarithm of resistivity of glass can in the first approximation be regarded as additive.

Although we have performed numerous calculations to check these formulas and the results appear encouraging, we do not believe that we have found the best variant of the method. Thorough verification of the experiment in practice is needed.

Literature Cited

1. E. A. Antonova, Candidate's dissertation, Leningr. Tekhnol. Inst. im. Lensoveta (1954).
2. M. S. Aslanova, Doctoral dissertation, Moscow (1955).
3. A. A. Appen and Kan Fu-hsi, Fiz. Tverd. Tela 1:1529 (1959).
4. A. A. Appen, Doctoral dissertation, Leningr. Tekhnol. Inst. im. Lensoveta (1952).
5. A. A. Appen and R. I. Bresker, Zh. Tekhn. Fiz. 22:946 (1952).

6. N. I. Brodskaya and V. S. Tatarinova, Uch. Zap. Leningr. Gos. Univ., Ser. Khim. Nauk (5):241 (1940).
7. L. I. Buneeva, et al., Use of Glasses for Insulator Production, TsBTI NIIÉP (1958).
8. O. K. Botvinkin, Collection on the Physics and Physical Chemistry of Glass, Gizlegprom (1933) p. 3.
9. V. N. Boricheva, Candidate's dissertation, Leningr. Tekhnol. Inst. im. Lensoveta (1956).
10. L. I. Buneeva, et al., Production of Glass Electrical Insulators, Moscow (1960).
11. N. P. Bogoroditskii and I. D. Fridberg, High-Frequency Inorganic Dielectrics, Izd. "Sovetskoe radio" (1948).
12. R. I. Bresker, Candidate's dissertation, Gos. Optich. Inst. (1949).
13. N. P. Bogoroditskii, V. V. Pasynkov, and B. M. Tareev, Electrotechnical Materials, Gosénergoizdat (1955).
14. N. P. Bogoroditskii, Izv. Tomsk. Politekhn. Inst., 91:299 (1956).
15. G. O. Bagoyk'yants, Collection: The Structure of Glass, Izd. Akad. Nauk SSSR (1955) p. 216. [English translation: The Structure of Glass, Vol. 1, Consultants Bureau, New York (1958), p. 167.]
16. G. M. Bartenev and A. S. Eremeeva, Kolloidn. Zh. 21:249 (1959).
17. N. P. Bogoroditskii and I. D. Fridberg, Zh. Tekhn. Fiz. 7:1905 (1937).
18. M. A. Bezborodov, et al., Phase Diagrams of Glassy Systems, Izd. Belorussk. Politekhn. Inst., Minsk (1959).
19. N. M. Verebeichik and V. I. Odelevskii, Collection: The Glassy State, Izd. Akad. Nauk SSSR (1960) p. 282. [English translation: The Structure of Glass, Vol. 2, Consultants Bureau, New York (1960), p. 248.]
20. V. V. Vargin and E. A. Antonova, Tr. Leningr. Tekhnol. Inst. im. Lensoveta (49):64 (1958).
21. V. V. Vargin, K. S. Evstrop'ev, et al., Physicochemical Properties of Glass and Their Dependence on Composition, Gizlegprom (1937).
22. A. G. Vlasov, Collection: The Glassy State, Izd. Akad. Nauk SSSR (1960) p. 222. [English translation: The Structure of Glass, Vol. 2, Consultants Bureau, New York (1960), p. 195.]
23. A. M. Venderovich and B. Lapkin, Zh. Éksperim. i Teor. Fiz. 9:46 (1939).
24. V. V. Vargin and E. A. Antonova, Tr. Leningr. Tekhnol. Inst. im. Lensoveta (49):55 (1958).
25. L. A. Grechanik, V. G Karpechenko, and N. V. Petrovykh, Scientific and Technical Collection of the Bureau of Technical Information of the Scientific Research Institute of Electrotechnical Glass, (14):29 (1959).
26. I. V Grebenshchikov and O. S. Molchanova, Zh. Obshch. Khim. 12:588 (1942).
27. N. A. Goryunova and B. T. Kolomiets, Zh. Tekhn. Fiz. 25:2069 (1955); Izv. Akad. Nauk SSSR, Ser. Fiz. 20:1496 (1956).
28. L. A. Grechanik, Zh. Prikl. Khim. 31:1164 (1958).
29. N. G. Gutkina, K. S. Evstrop'ev, and A. Ya. Kuznetsov, Zh. Tekhn. Fiz. 22: 1318 (1952).
30. B. M. Hochberg, Electrical Conductance of Dielectrics, Gostekhteoretizdat (1933).
31. M. A. Dolov, Uch. Zap. Kabardinsk. Gos. Ped. Inst. (10):13 (1956).
32. N. Ya. Dukarevich and N. T. Plashchinskii, Abstracts of Papers at the Third Interuniversity Conference on Modern Dielectric and Semiconductor Technology, Leningrad (1960) p. 35.
33. K. S. Evstrop'ev, A. Ya. Kuznetsov, and I. G. Mel'nikova, Zh. Tekhn. Fiz. 21:104 (1951).
34. K. K. Evstrop'ev and V. A. Khar'yuzov, Dokl. Akad. Nauk SSSR 136:140 (1961).
35. K. S. Evstrop'ev and A. O. Ivanov, Optiko-Mekh. Prom. (9):1 (1959).
36. K. S. Evstrop'ev and N. A. Toropov, Chemistry of Silicon and Physical Chemistry of Silicates, First ed., Fromstroiizdat (1950).
37. K. S. Evstrop'ev and N. A. Toropov, Chemistry of Silicon and Physical Chemistry of Silicates, Second ed., Promstroiizdat (1956).
38. K. S. Evstrop'ev, Tr. Leningr. Tekhnol. Inst. im. Lensoveta (45):54 (1958).
39. K. S. Evstrop'ev, Collection: The Structure of Glass, Izd. Akad. Nauk SSSR (1955) p. 9. [English translation: The Structure of Glass, Vol. 1, Consultants Bureau, New York (1958), p. 9.]
40. K. S. Evstrop'ev, O. V. Mazurin, and V. A. Khar'yuzov, Tr. Leningr. Tekhnol. Inst. im. Lensoveta (52:16 (1961).
41. K. S. Evstrop'ev and G. A. Pavlova, Tr. Leningr. Tekhn. Inst. im. Lensoveta (46:49 (1958).
42. K. K. Evstrop'ev, Collection: Physics of Dielectrics, Izd. Akad. Nauk SSSR (1960) p. 468.
43. S. P. Zhdanov and A. Ya. Kuznetsov, Dokl. Akad. Nauk SSSR 85:587 (1952).
44. S. P. Zhdanov, Doctoral dissertation, IKhS Akad. Nauk SSSR (1960).
45. L. A. Zhunina and N. A. Tinyakov, Izv. Vuzov. Énergetika (7):51 (1960).
46. A. F. Ioffe, The Physics of Crystals, Gosizdat (1929).

47. A. F. Ioffe, The Physics of Semiconductors, Izd. Akad. Nauk SSSR (1957).
48. V. A. Ioffe and Z. N. Zonn, Fiz. Tverd. Tela 3:1902 (1961).
49. V. A. Ioffe and G. I. Khvostenko, Fiz. Tverd. Tela 2:509 (1960).
50. V. A. Ioffe, Zh. Tekhn. Fiz. 24:661 (1954).
51. V. A. Ioffe, Zh. Tekhn. Fiz. 26:516 (1956).
52. V. A. Ioffe, Zh. Tekhn. Fiz. 27:1454 (1957).
53. V. A. Ioffe, I. V. Patrina, and S. V. Poberovskaya, Collection: The Glassy State, Izd. Akad. Nauk SSSR (1960) p. 454. [English translation: The Structure of Glass, Vol. 2. Consultants Bureau, New York (1960) p. 407.]
54. P. P. Kobeko, Amorphous Substances, Izd. Akad. Nauk SSSR (1952).
55. A. Ya. Kuznetsov and I. G. Mel'nikova, Zh. Tekhn. Fiz. 24:1204 (1950).
56. I. M. Kaplan and V. A. Chizhov, Steklo i Keram. (11):21 (1958).
57. B. T. Kolomiets, Zh. Vses. Khim. Obshchestva 5:553 (1960).
58. A. Ya. Kuznetsov and L. A. Pafomova, Optiko-Mekh. Prom. (11):1 (1959).
59. A. Ya. Kuznetsov, Doctoral dissertation, Gos. Optich. Inst. (1957).
60. A. Ya. Kuznetsov, Tr. Gos. Optich. Inst. 24(145):115 (1956).
61. L. Yu. Kurtts, Collection: Physicochemical Properties of the Ternary System, Izd. Akad. Nauk SSSR (1949) p. 110.
62. M. S. Kosman and N. N. Sozina, Zh. Éksperim. i Teor. Fiz. 17:341 (1947).
63. M. S. Kosman and R. T. Paranyuk, Zh. Éksperim. i Teor. Fiz. 24:721 (1953).
64. A. Ya. Kuznetsov, Zh. Fiz. Khim. 33:1492 (1959).
65. A. Ya. Kuznetsov, Zh. Fiz. Khim. 33:1726 (1959).
66. A. Ya. Kuznetsov, Optiko-Mekh. Prom. (5):35 (1960).
67. V. N. Kudin and M. S. Lapshin, Tr. Mosk. Énerg. Inst. (18):164 (1956).
68. B. T. Kolomiets and T. N. Vengel', Zh. Tekhn. Fiz. 27:2484 (1957).
69. M. A. Lebedinskii, Vacuum Tube Materials, Gosénergoizdat (1956).
70. M. L. Lyubimov, Metal—Glass Joints, Gosénergoizdat (1957).
71. M. S. Lapshin and B. M. Fradkin, Tr. Mosk. Énerg. Inst. (18):172 (1956).
72. I. G. Mel'nikova, K. S. Evstrop'ev, and A. Ya. Kuznetsov, Zh. Fiz. Khim. 25:1318 (1951).
73. O. V. Mazurin, Candidate's dissertation, Leningr. Tekhnol. Inst. im. Lensoveta (1953).
74. O. V. Mazurin, Tr. Leningr. Tekhnol. Inst. im Lensoveta (29):72 (1954).
75. O. V. Mazurin and E. V. Molchanov, Tr. Leningr. Tekhnol. Inst. im. Lensoveta (34):48 (1955).
76. O. V. Mazurin and E. S. Borisovskii, Zh. Tekhn. Fiz. 27:275 (1957).
77. O. V. Mazurin and G. T. Petrovskii, Tr. Leningr. Tekhnol. Inst. im. Lensoveta, Collection of students' papers, (1956) p. 30.
78. O. V. Mazurin and G. T. Petrovskii, Tr. Leningr. Tekhnol. Inst. im. Lensoveta, Collection of students' papers, (1956) p. 51.
79. O. V. Mazurin, É. V. Golikova, and N. V. Shtol'tser, Fiz. Tverd. Tela 1:630 (1959).
80. O. V. Mazurin and R. Z. Khomyakova, Tr. Leningr. Tekhnol. Inst. im. Lensoveta (43):72 (1957).
81. O. V. Mazurin, G. A. Pavlova, E. Ya. Lev, and E. K. Leko, Zh. Tekhn. Fiz. 27:2704 (1957).
82. O. V. Mazurin and A. S. Levin, Izv. Vuzov. Khimiya i Khim. Tekhnol. (2):142 (1958).
83. E. G. Momot, Radiotechnical Measurements, Gosénergoizdat (1957).
84. O. V. Mazurin and R. V. Brailovskaya, Fiz. Tverd. Tela 2:1477 (1960).
85. O. V. Mazurin and R. V. Brailovskaya, Dokl. Vyssh. Shkoly. Khimiya i Khim. Tekhnol. (2):383 (1959).
86. O. V. Mazurin, Collection: The Glassy State, Izd. Akad. Nauk SSSR (1960) p. 260. [English translation: The Structure of Glass, Vol. 2, Consultants Bureau, New York (1960), p. 229.]
87. I. G. Mel'nikova, A. Ya. Kuznetsov, and V. A. Brinberg, Zh. Fiz. Khim. 24:1294 (1950).
88. O. V. Mazurin and V. B. Brailovskii, Izv. Vuzov. Fiz. (1):117 (1959).
89. O. V. Mazurin and E. Ya. Lev, Izv. Vuzov. Fiz. (3):43 (1960).
90. M. A. Matveev and B. A. Kleimenov, Calculations in Glass Technology, Gizlegprom (1938).
91. O. V. Mazurin and V. A. Tsekhomskii, Tr. Leningr. Tekhnol. Inst. im. Lensoveta (59):36 (1961).
92. O. V. Mazurin and V. A. Tsekhomskii, Tr. Leningr. Tehnol. Inst. im. Lensoveta (59):33 (1961).
93. M. D. Mashkovich, Fiz. Tverd. Tela 3:1105 (1961).

94. O. V. Mazurin, Tr. Leningr. Tekhnol. Inst. im. Lensoveta (49):84 (1958).
95. T. M. Makarova, O. V. Mazurin, and V. S.Molchanov, Izv. Vuzov. Khimiya i Khim. Tekhnol. 3:1072 (1960).
96. B. I. Markin and R. L. Myuller, Zh. Fiz. Khim. 5:1262 (1934).
97. B. I. Markin and R. L. Myuller, Zh. Fiz. Khim. 7:592 (1936).
98. B. I. Markin, Zh. Tekhn. Fiz. 10:66 (1940).
99. B. I. Markin, Fiz. Tverd. Tela 3:450 (1961).
100. B. I. Markin, Zh. Fiz. Khim. 18:554 (1944).
101. B. I. Markin, Zh. Tekhn. Fiz. 22:932 (1952).
101. B. I. Markin, Doctoral dissertation, Leningr. Gos. Univ. im. Zhdanova (1954).
103. B. I. Markin, Zh. Obshch. Khim. 11:285 (1941).
104. B. I. Markin, Zh. Tekhn. Fiz. 22:941 (1952).
105. R. L. Myuller, Doctoral dissertation, Leningr. Gos. Univ. im. Zhdanova (1940).
106. R. L. Myuller, Zh. Tekhn. Fiz. 23:1875 (1953).
107. R. L. Myuller, Zh. Éksperim. i Teor. Fiz. 27:264 (1954).
108. R. L. Myuller, Zh. Tekhn. Fiz. 25:236 (1955).
109. R. L. Myuller, Zh. Tekhn. Fiz. 25:246 (1955).
110. R. L. Myuller, Zh. Tekhn. Fiz. 25:1556 (1955).
111. R. L. Myuller, Zh. Tekhn. Fiz. 25:1567 (1955).
112. R. L. Myuller, Zh. Tekhn. Fiz. 25:1868 (1955).
113. R. L. Myuller, Zh. Tekhn. Fiz. 25:2428 (1955).
114. R. L. Myuller, Zh. Tekhn. Fiz. 25:2440 (1955).
115. R. L. Myuller, Zh. Tekhn. Fiz. 26:2614 (1956).
116. R. L. Myuller and B. I. Markin, Zh. Fiz. Khim. 5:1271 (1934).
117. R. L. Myuller, Fiz. Tverd. Tela 2:1333, 1339, 1345 (1960).
118. R. L. Myuller, Summaries of Papers at the Third All-Union Conference on the Glassy State, Goskhimizdat (1959) p. 92.
119. A. V. Netushil, et al., High-Frequency Heating of Dielectrics and Semiconductors, Second ed., Gosénergoizdat (1959)
120. Scientific Literature on Dielectrics, Izd. Akad. Nauk SSSR (1952).
121. Scientific Literature on Dielectrics and Semiconductors, Bibliography for 1953-1955, Izd. LÉTI (1957).
122. Scientific Literature on Dielectrics and Semiconductors, Bibliography for 1956-1958, Izd. LÉTI (1960).
123. V. I. Odelevskii and R. N. Strel'tsyna, Izv. Tomsk. Politekhn. Inst. 91:323 (1956).
124. V. I. Odelevskii and N. M. Verebeichik, Izv. Tomsk. Politekhn. Inst. 91:247 (1956).
125. G. Ostroumov, Zh. Obshch. Khim. 19:407 (1949).
126. V. I. Odelevskii, N. M. Verebeichik, and L. M. Ped'ko, Collection: The Physics of Dielectrics, Izd. Akad. Nauk SSSR (1960) p. 170.
127. V. I. Odelevskii, Candidate's dissertation, Leningrad (1947); Zh. Tekhn. Fiz. 21:667 (1951).
128. G. T. Petrovskii, E. K. Leko, and O. V. Mazurin, Optiko-Mekh. Prom. (2):18 (1961).
129. V. P. Pryanishnikov, Collection: The Structure of Glass, Izd. Akad. Nauk SSSR (1955) p. 270. [English translation: The Structure of Glass, Vol. 1, Consultants Bureau, New York (1958), p. 214.]
130. A. I. Parfenov, A. F. Klimov, and O. V. Mazurin, Vestn. Leningr. Gos. Univ. (10):129 (1959).
131. V. P. Pryanishnikov, Quartz Glass, Promstroiizdat (1956).
132. V. P. Petrosyan, Candidate's dissertation, IKhS Akad. Nauk SSSR (1958).
133. E. A. Porai-Koshits, Collection: The Glassy State, Izd. Akad. Nauk SSSR (1960) p. 14. [English translation: The Structure of Glass, Vol. 2, Consultants Bureau, New York (1960), p. 9.]
134. N. M. Pavlushkin and G. G. Sentyurin, Manual of Glass Technology, Promstroiizdat (1957).
135. G. A. Pavlova, Tr. Leningr. Tekhnol. Inst. im. Lensoveta (46):56 (1958).
136. Committee of Technical Terminology of the Academy of Sciencies of the USSR, Recommended Terms, No. 53, Dielectrics, Izd. Akad. Nauk SSSR (1960).
137. G. I. Skanavi, The Physics of Dielectrics, Gostekhteoretizdat (1949).
138. G. I. Skanavi and A. M. Kashtanova, Zh. Tekhn. Fiz. 27:1770 (1957).

139. Collection: The Structure of Glass, Izd. Akad. Nauk SSSR (1955). [English translation: The Structure of Glass, Vol. 1, Consultants Bureau, New York (1958).]
140. G. I. Skanavi, Dielectric Polariazation and Losses in Glasses and Ceramic Materials with High Dielectric Constants, Gosénergoizdat (1952).
140a. J. M. Stevels, The Electrical Properties of Glass [Russian translation], IL (1961).
141. Collection: The Glassy State, Izd. Akad. Nauk SSSR (1960). [English translation: The Structure of Glass, Vol. 2, Consultants Bureau, New York (1960).]
142. Handbook of Electrotechnical Materials, Vol. 1, part 2, Gosénergoizdat (1960).
143. L. R. Takking and N. P. Shchegoleva, Uch. Zap. Leningr. Gos. Univ. Ser. Khim. Nauk (8):17 (1949).
144. V. V. Tarasov, New Aspects of the Physics of Glass, Gosstroiizdat (1959).
145. The Technology of Glass, edited by I. I. Kitaigorodskii, Third ed., Gosstroiizdat (1961).
146. Ya. I. Frenkel', Kinetic Theory of Liquids, Izd. Akad. Nauk SSSR (1945).
147. V. A. Florinskaya and R. S. Pechenkina, Collection: The Glassy State, Izd. Akad. Nauk SSR (1960) p. 157 [English translation: The Structure of Glass, Vol. 2, Consultants Bureau, New York (1960), p. 135]; V. A. Florinskaya, Collection: The Glassy State, Izd. Akad. Nauk SSSR (1960) p. 177 [English translation: ibid, p. 154].
148. Ya. I. Frenkel', Theory of Solids and Liquids, Gostekhteoretizdat (1934).
149. V. A. Khar'yuzov, Optiko-Mekh. Prom. (7):31 (1959).
150. V. A. Khar'yuzov, O. V. Mazurin, and N. M. Zubkova, Collection: The Glassy State, Izd. Akad. Nauk SSSR (1960) p. 264. [English translation: The Structure of Glass, Vol. 2, Consultants Bureau, New York (1960), p. 232.]
151. A. R. Hippel, Dielectric Materials and Applications [Russian translation], Gosénergoizdat (1959).
152. L. F. Shumitskaya, Collection of Technical Information, Central Scientific Research Laboratory for Electrical Glass, No. 3 (1955) p. 21.
153. V. I. Shelyubskii, Steklo i Keram. (9):13 (1953).
154. V. I. Shelyubskii, Zavodsk. Lab. 20:713 (1954).
155. S. A. Shchukarev and R. L. Myuller, Zh. Fiz. Khim. 1:625 (1930).
156. R. Ambronn, Ann. Phys. 58:139 (1919).
157. R. C. Burt, J. Opt. Soc. Am. 11:87 (1925).
158. A. E. Badger and J. E. White, J. Am. Ceram. Soc. 23:271 (1940).
159. H. Beyersdorfer, Silikattechnik 5:459 (1954).
160. F. Bischoff, Glastech. Ber. 28:98 (1955).
161. H. Buff, Liebigs Ann. 110:257 (1859).
162. E. P. Denton, H. Rawson, and J. E. Standworth, Nature 173:1030 (1954).
163. R. Douglas and J. Isard, J. Soc. Glass Technol. 33:289 (1949).
164. C. A. Elyard, P. L. Baynton, and H. Rawson, Glastech. Ber. 32K,(V(:36 (1959).
165. W. Eitel, The Physical Chemistry of the Silicates, Univ. of Chicago Press (1954).
166. W. Espe, Werkstoffkunde der Hochvakuumtechnik, Vol. II, Berlin (1960).
167. M. Fulda, Sprechsall (60):769, 789, 810, 831, 853 (1927).
168. M. Foex, Bull. Soc. chim. 11:456 (1944).
169. J. V. Fitzgerald, J. Am. Ceram. Soc. 34:314, 339; 389 (1951).
170. E. M. Guyer, J. Am. Ceram. Soc. 16:607 (1933).
171. G. Gehlhoff and M. Thomas, Z. Techn. Phys. 6:544 (1925).
172. R. L. Green, J. Am. Ceram. Soc. 25:83 (1942).
173. M. Gevers, Philips Res. Rep. 1:298, 447 (1946).
174. M. Gevers and F. K. du Pré, Trans. Faraday Soc. 42A:47 (1946).
175. E. M. Guyer, Proc. IRE 32:743 (1944).
176. W. Hinz and V. Havlicek, Glastech. Ber. 31:422 (1958).
177. W. E. Hauth and A. L. Pugh, Proc. 1956 Electron Components Sympos. New York, Engng. Publ. (1956) p. 37.
178. Y. Haven, Rec. Trav. Chim. 69:1259 (1950).
179. C. Hirayama and M. M. Rutter, J. Am. Ceram. Soc. 42:367 (1959).
179a. C. Hirayama and D. Berg, Am. Ceram. Soc. Bull. 40:551 (1961).

180. C. B. Hurd, E. W. Engel, and A. A. Vernon, J. Am. Chem. Soc. 49:417 (1927).
181. A. Heydweiller and F. Kopfermann, Ann. Phys. 32:739 (1910).
182. K. Hilbert, Silikattechnik 7:394 (1956).
183. L. Heroux, J. Appl. Phys. 29:1639 (1958).
184. A. Hippel, Dielectric Materials and Applications, New York (1954).
185. J. O. Isard, J. Soc. Glass Technol. 43:113 (1959).
186. B. Joyner and W. Bell, J. Am. Ceram. Soc. 36:263 (1953).
187. W. Jost, J. Chem. Phys. 1:466 (1933).
188. W. Kühne, Silikattechnik 7:451 (1956).
189. P. L. Kirby, Sci. Progr. 38:257 (1950).
190. C. A. Krauss and E. H. Darby, J. Am. Chem. Soc. 44:2783 (1923).
191. O. Kubaschewski, Z. Elektrochem. 42:5 (1936).
192. R. Kamel, J. Appl. Phys. 24:1308 (1953).
193. J. T. Littleton and G. W. Morey, The Electrical Properties of Glass, New York (1933).
194. J. T. Littleton and W. L. Wetmore, J. Am. Ceram. Soc. 19:243 (1936).
195. B. Lengyel, Glastech. Ber. 18:177 (1940).
196. B. Lengyel and Z. Boksay, Z. Physik Chem. 203:93 (1954).
197. B. Lengyel and Z. Boksay, Z. Physik Chem. 204:157 (1955).
198. B. Lengyel, M. Somogyi, and Z. Boksay, Z. Physik Chem. 209:15 (1958).
199. B. Lenbyel, Z. Boskaya, and F. Gallyas, Magy. Tud. Akad. Kém. Tud. Oszt. Közlomen. 15:35 (1961).
200. H. Moore and R. De-Silva, J. Soc. Glass Tehnol.36:5 (1952).
201. I. Mohyuddin and R. Douglas, Physics and Chem. of Glasses (1):71 (1960).
202. G. Morey, The Properties of Glass, Second ed., New York (1954).
203. G. Morey, The Properties of Glass, New York (1938).
204. N. F. Mott and J. T. Littleton, Trans. Faraday Soc. 34:485 (1938).
205. L. S. McDowell and H. L. Begeman, Phys. Rev. 33:55 (1929).
206. L. Navias and R. L. Green, J. Am. Ceram. Soc. 29:267 (1946).
206a. A. E. Owen, Physics and Chem. of Glasses 2:87 (1961).
207. G. Petrovskij, Silikaty 3:336 (1959).
207a. L. Prod'homme, Verres et Réfractaires 14:69, 124 (1960).
208. F. Quittner, Ann. Phys. 85:745 (1928).
209. H. Rötger, Glastech. Ber. 19:192 (1941).
210. H. Rötger, Glastech. Ber. 31:54 (1958).
211. D. M. Robinson, Physics 2:52 (1932).
212. W. Rothe, Silikattechn. 12:9 (1961).
213. H. Rötger, Wiss. Z. Friedrich-Schiller-Univ. Jena 7, Math.-Naturwiss. Reihe, (2/3):237 (1957/58).
214. G. Schulze, Ann. Phys. 37:435 (1912).
215. E. Seddon, E. Tippet, and W. E. S. Turner, J. Soc. Glass Technol. 16:450 (1932).
216. R. Schwarz and I. Halberstadt, Z. anorg. Chem. 199:33 (1931).
217. H. Siemen, Phys. Rev. 31:119 (1928).
218. R. B. Sosman, Properties of Silica, New York (1927).
219. E. R. Schweidler, Ann. Phys. 24:711 (1907).
220. W. Schottky, Z. Physik Chem. 29:335 (1935).
220a. P. M. Sutton, Progr. in Dielectrics, Vol. 2, London (1960) p. 113.
221. J. M. Stevels, Philips Techn. Rev. 13:360 (1952).
222. J. M. Stevels, Progress in the Theory of the Physical Properties of Glass, New York (1948).
223. J. M. Stevels, Handbuch der Physik, Vol. XX, Chapter III, "The Electrical Properties of Glass," Springer-Verlag, Berlin (1957).
224. M. Strutt, Archiv Electrotechn. 25:715 (1931).
225. J. M. Stevels, J. Soc. Glass Technol. 34:80 (1950).
226. S. W. Strauss and others, J. Res. Nat. Bur. Std. 56:135 (1954).
227. J. E. Standworth, J. Soc. Glass Technol. 30:381 (1946).

228. J. M. Stevels, Philips Res. Rep. 6:34 (1951).
229. M. Strutt, vand der Ziel, Physica 10:445 (1943).
230. J. H. Terry, Ceramic Ind. 74:100 (1960); 75:44, 51 (1960).
231. F. Tegetmeier, Ann. Phys. Chem. 41:18 (1890).
232. F. Tank, Ann. Phys. 48:307 (1915).
233. H. E.Taylor, Trans. Faraday Soc. 52:873 (1956).
234. R. Ure, J. Chem. Phys. 26:1363 (1957).
235. J. Vermeer, Physica 22:1257 (1956).
236. J. Volger and J. M Stevels, Philips Res. Rep. 11:452 (1956).
237. J. Volger, Bull. Inst. Intern. Froid. Annexe (2):89 (1955).
238. J. Volger, M. J. Stevels, and C. van Amerongen, Philips Res. Rep. 8 452 (1953).
239. A. E. Williams, Chem. Age 53:97 (1945).
240. C. Wagner and W. Schottky, Z. Physik Chem. 11:163 (1930).
241. E. Warburg, Ann. Phys. Chem. 21:622 (1884).
242. E. Warburg and F. Tegetmeier, Ann. Phys. Chem. 35:455 (1888).
243. V. Zworykin, Phys. Rev. 27:813 (1926).
244. N. S. Andreev, V. I. Aver'yanov, and E. A. Porai-Koshits, Collection: Structural Transformations in Glasses at High Temperatures, Izd. Nauka, Moscow–Leningrad (1965), p. 59. [English translation: The Structure of Glass, Vol. 5, Consultants Bureau, New York (1965), p. 49.]
245. V. B. Brailovskii, Author's abstract of candidate's dissertation, Lensovet Leningrad Technical Institute (1965).
246. N. P. Bogoroditskii and N. D. Fridberg, Fiz. Tverd. Tela 6: 680 (1964).
247. V. B. Brailovskii, M. Ya. Rozenblyum, and O. V. Mazurin, Collection: Electrical Properties and the Structure of Glass, Izd. Khimiya, Moscow–Leningrad (1964), p. 63. [English translation: this volume, p. 102.]
248. D. A. Goganov and E. A. Porai-Koshits, Collection: Structural Transformations in Glasses at High Temperatures, Izd. Nauka, Moscow–Leningrad (1965), p. 100. [English translation: The Structure of Glass, Vol. 5, Consultants Bureau, New York (1965), p. 82.]
249. L. A. Grechanik, E. A. Fainberg, and I. N. Zertsalova, Zh. Prikl. Khim. 36: 91 (1963).
250. K. S. Evstrop'ev, Collection: Electrical Properties and the Structure of Glass, Izd. Khimiya, Moscow–Leningrad (1964), p. 7. [English translation: this volume, p. 59.]
251. K. K. Evstrop'ev, Author's abstract of candidate's dissertation, A. A. Zhdanova, Leningrad State University (1962).
252. I. N. Zertsalova, Collection: Electrical Properties and the Structure of Glass, Izd. Khimiya, Moscow–Leningrad (1964), p. 112. [English translation: this volume, p. 141.]
253. A. P. Zorin and O. V. Mazurin, Collection: Electrical Properties and the Structure of Glass, Izd. Khimiya, Moscow–Leningrad (1964), p. 65. [English translation: this volume, p. 104.]
254. V. A. Ioffe et al., (Discussion), Collection: Electrical Properties and the Structure of Glass, Izd. Khimiya, Moscow–Leningrad (1964), p. 93. [English translation: this volume. p. 124.]
255. A. O. Ivanov, K. S. Evstrop'ev, and M. L. Dorokhova, Collection: Electrical Properties and the Structure of Glass, Izd. Khimiya, Moscow–Leningrad (1964), p. 47. [English translation: this volume, p. 86.]
256. A. O. Ivanov, Fiz. Tverd. Tela 5: 2447 (1963).
257. A. I. Ivanov and E. I. Galant, Collection: Electrical Properties and the Structure of Glass, Izd. Khimiya, Moscow–Leningrad (1964), p. 44. [English translation: this volume, p. 84.]
258. A. Ya. Kuznetsov and V. A. Tsekhomskii, Collection: Electrical Properties and the Structure of Glass, Izd. Khimiya, Moscow–Leningrad (1964), p. 105. [English translation: this volume, p. 136.]
259. O. V. Mazurin, Collection: Electrical Properties and the Structure of Glass, Izd. Khimiya, Moscow–Leningrad (1964), p. 22. [English translation: this volume, p. 69.]
260. O. V. Mazurin and V. A. Tsekhomskii, Izv. Vuzov. Fizika, No. 1: 125 (1964).
261. O. V. Mazurin, Author's abstract of doctoral dissertation, Lensovet Leningrad Technical Institute (1962).
262. O. V. Mazurin, Electrical Properties of Glass, Khimizdat, Leningrad (1962), Chapter III.
263. R. L. Myuller and A. A. Pronkin, Collection: Electrical Properties and the Structure of Glass, Izd. Khimiya, Moscow–Leningrad (1964), p. 51. [English translation: this volume, p. 93.]

264. G. A. Pavlova, Collection: Electrical Properties and the Structure of Glass, Izd. Khimiya, Moscow–Leningrad (1964), p. 81. [English translation: this volume, p. 116.]
265. A. A. Pronkin, Zh. Prikl. Khim. 37: 887 (1964).
266. V. A. Tsekhomskii, O. V. Mazurin, and K. K. Evstrop'ev, Fiz. Tverd. Tela 5: 586 (1963).
267. P. J. Bray and J. G. O'Keefe, Phys. Chem. Glasses 4: 37 (1963).
268. R. J. Charles, J. Am. Ceram. Soc. 46: 235 (1963).
269. Ch. Hirayama and D. Berg, J. Am. Ceram. Soc. 46: 85 (1963).
270. B. Lengyel and Z. Boksay, Z. Physik. Chem. 222: 183 (1963).
271. B. Lengyel and Z. Boksay, Z. Physik. Chem. 223: 49 (1963).
272. B. Lengyel, Z. Boksay, and S. Dobos, Z. Physik. Chem. 223: 186 (1963).
273. G. Milnes and J. Isard, Phys. Chem. Glasses 3: 157 (1962).
274. A. E. Owen, Progress in Ceramic Science, Vol. 3, Pergamon Press, Oxford (1963), p. 78.
275. C. Sella, T. -L. Tran, M. Navez, and J. J. Trillat, Silicates Ind. 29: 15 (1964).
276. T. -L. Tran and C. Sella, Compt. Rend. 258: 4234 (1964).

I

GENERAL ASPECTS OF THE STUDY OF ELECTRICAL PROPERTIES OF GLASSES

HISTORY OF RESEARCH ON THE ELECTRICAL PROPERTIES OF GLASSES

K. S. Evstrop'ev

The history of 40 years of research on the electrical properties of glass in the USSR represents a valuable contribution to the physical chemistry and technology of glass.

1. The remarkable investigations of crystalline dielectrics, started by A. F. Ioffe [1] in the Institute of Technical Physics, were successfully continued in glass research at the A. A. Zhdanov Leningrad State University, the S. I. Vavilov State Optical Institute, the Lensoviet Technological Institute in Leningrad, etc. Special credit is due to Ya. I. Frenkel' [2]; his work on the theory of solid (crystalline) and liquid substances was the foundation of systematic investigations of the effect of temperature on the electrical conductivity of solid glasses. The applicability of the Frenkel' exponential $\varkappa = Ae^{-\alpha/kT}$ to solid glasses was proved theoretically by R. L. Myuller [3]. His interpretation of the exponential itself and of the preexponential is especially interesting.

R. L. Myuller formulated certain general propositions concerning the structure of inorganic glassy systems. They related primarily to the role of directed covalent bonds in the creation of a glassy structure and in the formation of complexes by certain structural elements. In this series of investigations, the concepts of the dual position of alkali-metal ions in the structure of glass (electrically undissociated, O^-Na^+, and dissociated, $\overset{|}{\underset{|}{O}}\,Na^+$), the mechanism of ionic movement in the interstices, and the presence of "differentiated groupings of homogeneous glass components" were put forward on the basis of conductivity data on simple two- and three-component borate glasses.

Recently R. L. Myuller, as the result of evaluation of experimental data on the physicochemical properties of chalcogenide glasses, again concluded that differentiated amicronic formations exist in their structure.

2. Analysis of conductivity versus composition diagrams of molten binary glassy systems, studied by numerous investigators, indicates that they can be represented in the first approximation (especially in the high-temperature region) by the exponential expression

$$\log \varkappa = ac + b$$

where c is the concentration of the variable component; a and b are constants.

However, this is valid only when the phase diagram of the system corresponds to a simple eutectic. The log $\varkappa$ versus c relationship for systems forming continuous series of solid solutions is curvilinear. In a eutectic system with a definite compound the log $\varkappa$ versus c plot consists of two straight lines intersecting at a point corresponding to the composition of that compound [4].

If this view is accepted as a first approximation, it may be assumed that the additivity of log $\varkappa$ is evidence of the additivity of the activation energy of conduction, while this in turn suggests the existence of additivity of structural formation in melts, i.e., the possible existence of amicronic heterogeneities in them. This should be regarded as merely an approximation, as a scheme which often becomes more complicated; the extent of the complication increases with fall of the melt temperature, especially as the melt passes first into the softened and then into the solid state [5].

3. It is known that the bivalent-metal oxides BeO, MgO CaO, SrO, BaO, PbO, and ZnO, when introduced into sodium silicate glasses, lower the conductivity to an increasing extent as the effective radius of the metal ion increases [6]. These observations indicate that the fall of conductivity can be due only to the drag exerted on the Na ion by the oxygen component of the glass under these conditions. We know that the energy and structural state of oxygen in glass, as measured, for example, by its refraction, varies over a fairly wide range in accordance with the composition of the glass and especially with the contents of other ions.

The different refractive states of oxygen in glass determine also the force of its interaction with dissociated Na atoms, and hence the influence of the energy state of oxygen on the conductivity.

4. R. I Bresker found for alkali borate glasses, and later A. O. Ivanov also found for alkali germanate glasses [7], that the electrical properties of these glasses are influenced by changes in the coordination of B and Ge, respectively. It was found that introduction of oxygen together with Na ions into B_2O_3 or GeO_2, up to a certain Na concentration (about 15%), results in the formation of groups with larger coordination numbers (4 for B and 6 for Ge), which causes more stable bonding between the Na ions and the coordination groups and decreases the diffusion coefficient of the Na ion. In silicate glasses, on the other hand, the Na_2O introduced into the glass interacts with $-\overset{|}{\underset{|}{Si}}-O-\overset{|}{\underset{|}{Si}}-$ and, breaking Si–O–Si bonds, loosens the structural network and raises the conductivity of the glass.

In borate and germanate glasses further increase of the Na_2O concentration produces a considerable increase of conductivity if contacts between the coordinated structural groups (tetrahedrons or octahedrons) become possible.

Similar results were obtained by A. O. Ivanov and E. I. Galant [8] in a study of the electrical conductivity of certain gallium glasses.

5. The effects of heat treatment, chilling and annealing are of great importance in investigations of the electrical properties of glasses. It is known that chilling increases the conductivity of glasses whereas annealing raises the resistance.

These facts indicate that chilling preserves to some extent the structural state of glass at higher temperatures and thus determines the higher conductivity.

However, there have been no systematic studies up to now of the effect of heat treatment on the electrical properties of glasses.

Z. A. Levtsova [9] studied the effects of chilling and annealing on the conductivity of glasses and discovered some new relationships applicable both to glasses containing alkaline oxides and to alkali-free glasses. It was assumed that in conductivity measurements on ordinary alkali glasses and on alkali-free glasses with conductivities not below 10^{-14} $ohm^{-1} \cdot cm^{-1}$ the influence of surface conductivity due to the presence of adsorbed water on the specimen surface should be taken into account only at temperatures below 100°C; at higher temperatures it may be disregarded.

It was found that in studies of the effect of chilling on alkali-free glasses of high resistivity the influence of surface conductivity must be taken into account even at high temperatures (above 100-150°C). For such glasses the influence of surface conductivity becomes negligible only at temperatures above 400-450°C.

The effects of chilling and annealing were studied in a large group of optical glasses of various chemical compositions, including both alkali and alkali-free glasses: K8, K14, KF4, LF6, F4, and TF1 (alkali glasses), and TK3 and BF16 (alkali-free).

It was confirmed that for alkali glasses the conductivity variations follow the same course as the degree of chilling assessed from the birefringence.

The effects of chilling differ for different glasses:

a) In one group of glasses, the log ρ versus $1/T$ plots for annealed and chilled specimens are parallel,

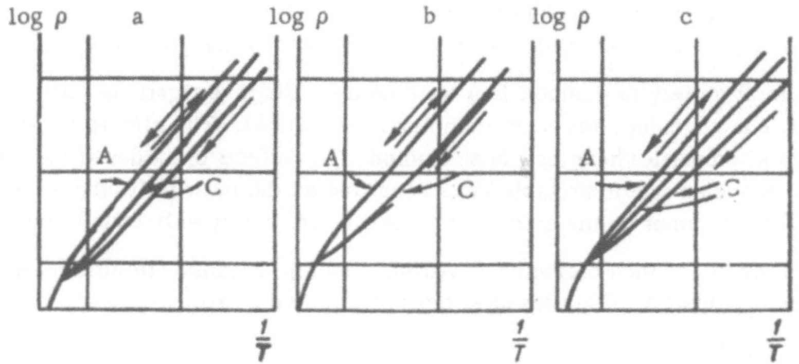

Effects of chilling on the variations of resistivity of different glasses with temperature. A) Annealed; C) chilled specimens.

and the distance between them is proportional to the degree of chilling; in the annealing range the two branches curve and converge (see the figure, a).

b) In another group, the conductivity – temperature relationships for chilled specimens form a system of straight lines intersecting in pairs at one or two points (or a single curve), while for annealed specimens the plot of log ρ versus $1/T$ is a straight line with a bend in the annealing region (see the figure, b).

c) A third type of effect produced by chilling is represented by turning of the log ρ versus $1/T$ plot through a small angle about a point in the annealing range. The angle through which the linear plot turns is a measure of the degree of chilling (see the figure, c).

d) In the last case chilling has no appreciable influence on conductivity. This was found for lead silicate glasses, which exhibited a considerable birefringence effect without any apparent change of conductivity.

6. If we examine the results of researches carried out in the USSR on the electrical properties of glasses, we can claim with confidence that many problems covering the most important purely scientific, theoretical, and applied aspects have been fully studied.

We will enumerate only some of them. Practical problems have included the following:

1. Development of highly insulating and special glasses for the electrical and vacuum tube industries.
2. Development of methods for electrical melting of glasses with the use of Joule heat and of electric and magnetic losses.
3. Development of techniques for raising and lowering the surface conductivity of glasses, and practical applications of these techniques in industry.
4. Investigation of the electrical properties of glass fibers and fabrics.
5. Investigation of the effects of glass composition on the magnitude of the glass–solution phase potential, development of the theory and practice of the glass electrode, etc.

Many institutes and factories took part in studies of these important problems, which are of great importance to the national economy.

Some of the most important scientific problems which have been and are being solved by Soviet scientists are listed below:

1. Study of the laws governing the effects of temperature on conductivity of solid and molten glasses.
2. Discovery of the basic laws of ionic conductivity of glasses.
3. Detailed study and interpretation of the polyalkali effect.
4. Investigation of transference numbers in solid glasses.
5. Study of the polarization effect when electricity passes through glass.
6. Discovery and study of semiconductor conductivity in glasses.
7. Study of the electric strength of glasses.

8. Ion diffusion and conductivity of glasses.
9. Electrical properties of crystalline glass materials, and many other problems.

Glasses of a great variety of compositions were used for the investigations: silicate, borate, germanate, phosphate, fluoride, chalcogenide, and more complex compositions. The effects of the oxides of very many elements on electrical properties have now been studied. The effects of oxides of the elements in the top left-hand corner of the Mendeleev periodic table have been studied the most fully; the effects of oxides of elements in the bottom right-hand corner on the electrical properties are not as well-understood.

The effects of oxides of such elements as yttrium, indium, hafnium, tellurium, gold, palladium, platinum, etc., on the electrical properties of glasses have not been studied at all.

This is a task for the future.

7. At the Sixth International Congress on Glass, held in Washington in July, 1962, of the 105 papers presented only five were concerned with the electrical properties of glasses, including one communication in connection with 25 years of operation of electrical tank furnaces with graphite electrodes. The communication on electrical melting was by Borel (Switzerland) and Touvay. The remaining papers are concerned with studies of the electrical properties of certain glasses and provide reference material.

The paper by McMillan (England), entitled "The preparation and properties of glasses containing major proportions of certain transition-metal oxides," describes the properties of three-component aluminate glasses containing MnO, CoO, or FeO.

A brief communication by Chose and Goug-Jen Su (Rochester University, New York) refers to the viscosity, density, and electrical resistance of ternary lanthanum borate melts containing oxides of barium, tantalum, or tungsten. The determinations were performed at 1000-1265°C.

Stevels and Trap, in their communication "New types of glasses showing electronic conductivity," state facts which are well known to us. They note that inorganic oxide systems can acquire some degree of electronic conductivity after introduction of variable-valence elements.

An examination of the Congress papers concerned with the electrical properties of glasses leads to the conclusion that nothing essentially new or especially interesting, in either the experimental or the theoretical sense, was reported. Similar results for the corresponding glassy systems are already known from the literature, mainly from Soviet work.

Of other Congress papers having some bearing on the electrical properties of glasses, mention must be made of the paper by Stookey, Olcott, et al.: "Ultrahigh-strength glasses by ion exchange and surface crystallization."

8. One defect in research on the electrical properties of glasses is that relatively little is known about the conductivity and dielectric properties of glassy melts. At one time this gap was due to purely practical difficulties in conductivity determination techniques and to the lower practical significance of conductivity at high temperatures. Neither cause is now valid, because methods ensuring the required accuracy have been devised, while development of electrical methods of glass melting is undoubtedly connected with the electrical properties of melts.

The course of structural changes during solidification of melts can be revealed by investigations of the variations of conductivity with temperature in the high-temperature range in conjunction with studies of the effect of chemical composition on the conductivity of glassy melts.

Conclusions regarding the structure of glasses, drawn from the results of investigations of their properties in the solid state only and not verified for softened and glassy melts, always leave a number of unsolved problems. Nothing definite is known about transference numbers or diffusion of ions in melts, or the changes which directed covalent bonds coordination groups, and the initial structural groups undergo during softening in the melt itself. Until these problems have been solved it is difficult to study the electrical properties of crystalline glass materials and the nucleation effects in their formation, semiconductor materials, and many other problems of modern technology.

There has been little research on the dielectric strength of glasses, or on its dependence on temperature, chemical composition, heat treatment, and structural changes. Little is known about the effects of crystallization and nucleation on the dielectric strength of crystalline glass materials.

9. The wealth of experimental data obtained in studies of the electrical properties of glasses in all their aspects requires appropriate generalization, additional theoretical evaluation, and publication in the form of a large monograph on the electrical properties of glasses. This monograph should be written by Soviet scientists, as their work has made the most systematic, diverse, and complete contributions in this field.

Literature Cited

1. A. F. Ioffe, Physics of Crystals, GIZ (1929).
2. Ya. I. Frenkel', Theory of Solids and Liquids, Gosteoretizdat (1934).
3. R. L. Myuller, Izv. Akad. Nauk SSSR, Ser. Fiz. IV:607 (1940); Author's summary of doctoral dissertation, Leningr. Gos. Univ. im. A. A. Zhdanova (1940); Collection: The Glassy State, Izd. Akad. Nauk SSSR (1960) p. 61. [English translation: The Structure of Glass, Vol. 2, Consultants Bureau, New York (1960) p. 50.]
4. K. S. Evstrop'ev, Izv. Akad. Nauk SSSR, Ser. Fiz. IV:616 (1940).
5. K. S. Evstrop'ev, Collection: The Glassy State, Izd. Akad. Nauk SSSR (1960) p. 39. [English translation: The Structure of Glass, Vol. 2, Consultants Bureau, New York (1960) p. 32.]
6. O. V. Mazurin, "Electrical properties of glass," Tr. Leningr. Tekhnol. Inst. im. Lensoveta, No. 62, Goskhimizdat (1962).
7. A. O. Ivanov and K. S. Evstrop'ev, Dokl. Akad. Nauk SSSR, 145:797 (1962).
8. A. O. Ivanov and E. I. Galant, this collection, p. 84.
9. Z. A. Levtsova, Tr. Leningr. Tekhnol. Inst. im.Lensoveta (in press).
10. Advances in Glass Technology, Plenum Press, New York (1962).

THERMAL IONIZATION IN GLASSES AND MOBILITY OF CURRENT CARRIERS IN THEM

R. L. Myuller

Glassy substances are characterized by the presence of a spatial framework of covalently bonded atoms; for example, $(SiO_{4/2})_n$ in the case of silica [1]. This framework confers mechanical strength and chemical durability but at the same time separates, and therefore weakens the interaction of, polar groups consisting of ionized oxygen, boron, or aluminum atoms and of metal cations joined directly to them by Coulomb forces ($\overset{+}{M}\overset{-}{O}SiO_{3/2}$, $\overset{+}{M}\overset{-}{B}O_{4/2}$, $M^{+}Al^{-}O_{4/2}$). In contrast to typical ionic crystals, in these glassy substances short-range covalent bond forces predominate; these forces determine the short-range atomic order in such structural elements as $SiO_{4/2}$, $BO_{3/2}$, $B^{-}O_{4/2}$, etc.

The absence of well-defined long-range Coulomb interaction forces explains why the degree of geometrical order in the structure of these substances has little influence on the energy. This is illustrated by the data in Table 1. This gives the "atomization" (dissociation) energies of a number of glass-forming substances in the crystalline and glassy states. These data were obtained from the standard enthalpies of the solids at 25°C and the heats of sublimation [2, 3, 4] in accordance with the equation

$$\Delta H^0_{298}(xA, yB, zC\ldots)_A = -\Delta H^0_{298}(A_xB_yC_z\ldots)_{solid} + x\,\Delta H^0_{298}(A)_{gas} + y\,\Delta H^0_{298}(B)_{gas} + z\,\Delta H^0_{298}(C)_{gas} + \ldots$$

Comparison of the dissociation energies of these solids in the crystalline and glassy states shows that the values differ little even when the content of polar structural elements is considerable. Since long-range geometrical order makes no significant contribution to the dissociation energy, the latter can, in the first approximation,

TABLE 1. Dissociation Energies of Solids in the Glassy and Various Crystalline States

Chemical composition	Dissociation energy, kcal/mole			
	Crystals		Glass	Difference,%
SiO_2	Quartz	411.8	408.9	0.7
	Cristobalite	411.4		
	Tridymite	411.2		
B_2O_3		674.0	669.6	0.7
P_4O_{10}		1612.8	1626.8	0.9
Na_2SiO_3		593.0	590.0	0.5
$K_2O \cdot Al_2O_3 \cdot 4SiO_2$	Leucite	2662.0	2650.6	0.4
$K_2O \cdot Al_2O_3 \cdot 6SiO_2$	Microcline	3484.0	3448.0	1.0
	Adularite	3510.0		1.8
$CaO \cdot 2B_2O_3$		1650.1	1637.5	0.8
$12CaO \cdot 7Al_2O_3$		4617.0	4585.0	0.7
$2CaO \cdot Al_2O_3$		704.0	695.0	1.3
$3CaO \cdot Al_2O_3$		861.0	848.0	1.5
Se		49.4	48,3	2.3

TABLE 2. Chemical Bond Energies in Glassy Solids from Dissociation Energy Data

Chemical composition	Structural unit	Dissociation energy, kcal/mole	Chemical bond	Bond energy, kcal/mole
SiO_2	$(SiO_{4/2})$	409	Si—O	102
B_2O_3	$(BO_{3/2})$	335	B—O	112
P_4O_{10}	$(O^-P^+O_{3/2})$	407	P—O	81
Na_2SiO_3	$(Na_2^+O_{2/2}^-)(SiO_{4/2})$	590	Na^+O^-	91
$K_2O \cdot Al_2O_3 \cdot 4SiO_2$	$(K^+Al^-O_{4/2})_2(SiO_{4/2})_4$	2650	K^+Al^-	98
$K_2O \cdot Al_2O_3 \cdot 6SiO_2$	$(K^+Al^-O_{4/2})_2(SiO_{4/2})_6$	3448	K^+Al^-	88
S	$(SS_{2/2})$	—	S—S	55 [5]
Se	$(SeSe_{2/2})$	94	Se—Se	47
As_2S_3	$(AsS_{3/2})$	—	As—S	51 [5]
As_2Se_3	$(AsSe_{3/2})$	—	As—Se	45 [6]
$GeSe_2$	$(GeSe_{4/2})$	—	Ge—Se	49 [6]

be assigned to "short-range" atomic interaction in the structural units $SiO_{4/2}$, $BO_{3/2}$, $AlO_{3/2}$, etc. If the dissociation energy is thus ascribed primarily to localized chemical bonds, the energies required to break the corresponding single chemical bonds can be estimated [4]. Table 2 gives the energies of localized single chemical bonds determined in this manner. It was assumed that the bond energies in structural $SiO_{4/2}$ and $Al^-O_{4/2}$ are similar (approximately 102 kcal/mole of bonds).

It is of interest to compare these thermochemically determined values of the energies of covalent and virtually homopolar [11] bonds with their ionization energies determined from the boundaries of light absorption and from the conductivity – temperature relationships for chalcogenide glasses in the system Ga–Ge–As–Se–S under conditions of through conduction ($\log \beta \approx 0$) [1, 9]. This comparison is made in Table 3 from experimental data on the conductivities of glassy Se [9], $AsS_{2.5}$ [5], $AsSe_{1.5}$ [7, 9], $GeSe_4$, and $AsGe_{1.5}Se_{4.5}$ [6, 10]. Evidently under these conditions of free through conduction the ionization energy of the valence bonds in these glasses is close to the energy of these bonds and is in the range of 1.7-2.2 eV. It also follows that the activation energy of translational displacement of free current carriers (electrons and holes) $\varepsilon_a = 0.5(\varepsilon_\sigma - \varepsilon_\lambda)$ [9] does not exceed 0.2 eV and is about 10% of the ionization energy.

TABLE 3. Ionization Energies of Covalent Bonds and the Activation Energies for Displacement of Free Current Carriers in Chalcogenide Semiconductor Glasses

Chemical bond	Bond energy, eV	Boundary of light absorption ϵ_λ, eV	Energy of conduction ϵ_σ eV	Activation energy ϵ_a, eV
Se—Se	2.0	1.76 [7, 8]	1.7 [8, 9]	0.05
As—S	2.2	2.2 [5]	2.2 [5]	0.05
As—Se	2.0	1.7 [7, 9]	1.7 [7, 9]	0.05
Ge—Se	2.1	1.8 [6, 10]	2.2 [6, 10]	0.2

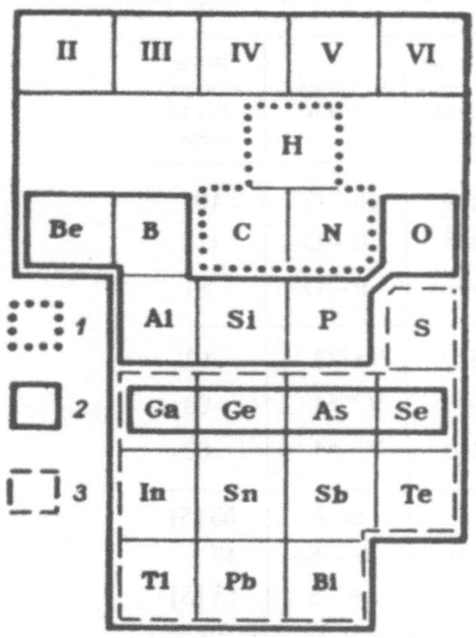

Elements of the principal subgroups of the Mendeleev periodic system, forming organic polymers (1) and inorganic glassy polymers: ionic oxide type (2) and oxygen-free semiconductor type (3).

The elements of principal subgroups III-IV of the Mendeleev periodic system, which form semiconducting glasses, are directly adjacent to the elements B, Al, Si, P, O, which form the basis of inorganic oxide glasses (see the figure). These glasses have high valence-bond ionization energies, greater than 3.5 eV according to Table 2. The electronic conductivity of such glasses is overlapped by ionic conductivity, with values of about 1-2 eV for the energy of electrolytic dissociation of the polar M^+X^- groups. Although the energy of rupture of the ionic bonds in these polar groups is about 4 eV (Table 2) and differs little from the ionization energy of covalent bonds, nevertheless the energy of electrolytic dissociation is low (about 1-2 eV). This is the consequence of the high value of the quasisolvation energy effect of the interaction of the free cations formed by dissociation and the negatively charged holes with the surrounding dipoles of the polar groups [12]. Dipole orientation is effected by displacement of the cations within the polar structural units [13]. Calculations by the Frenkel' – Onsager method show that in alkali borate glasses with a dielectric constant of about 18 the energy required to remove a cation from a polar $Na^+B^-O_{4/2}$ structural unit is overlapped by 85% by the energy of the quasisolvation effect [12]. In this case the electrolytic dissociation energy ε_i falls to 2.1 eV and the activation energy ε_a of translational displacement of the free cation is about 0.2 eV.

Thus, both in electronically conducting and in ionically conducting glasses the activation energies of displacement of free current carriers are small in comparison with the ionization energies of the covalent bonds or with the electrolytic dissociation energies of the polar units. The value of ε_σ ("energy of conduction") derived from the conductivity – temperature relationship of glasses is determined in the first approximation by the ionization energy ε_i of the structural element which establishes the concentration of the current carriers. For example, when $\varepsilon_i = 30$ kcal/mole and $\varepsilon_a = 5$ kcal/mole we have, at 500°K, the value of $\alpha_i = \exp(-\varepsilon_i/2kT) \approx 10^{-7}$ for the proportion of ionized structural elements among the total number of "free" carriers, and $\alpha_a = \exp(-\varepsilon_a/kT) \approx 10^{-2}$ for the proportion of current carriers in translational motion. In other words, $\alpha_i \approx 10^{-5}\alpha_a$.

We know that introduction of oxides of transition elements with d-electrons into low-alkali oxide glasses is accompanied by a sharp decrease in the formation energy of free electrons ($\varepsilon_i < 0.5$ eV). In such glasses the lower ionic conductivity is easily overlapped by electronic conductivity.

The relationships noted here for oxide and oxygen-free inorganic glasses should also apply to organic polymers. The latter are immediately adjacent in chemical composition and properties to inorganic polymers; inorganic glass-forming substances are essentially of this type (see the figure).

The complete electrolytic dissociation of Na^+O^- polar groups in the molten silicate glass investigated by A. F. Borisov and V. I. Zadumin [14] appears improbable in the light of the foregoing. From their values for

TABLE 4

Structural element	Conc. of free current carriers		Increment per 100°
	at 1500°K	at 1600°K	
Na^+O^-	$0.020\,n$	$0.025\,n$	$0.005\,n$
Si—O	$3.2 \cdot 10^{-6}\,n$	$6.3 \cdot 10^{-6}\,n$	$3 \cdot 10^{-6}\,n$

the bond energies of Si–O (75 kcal/mole) and Na^+–O^- (23 kcal/mole) the values given in Table 4 are to be expected for the concentrations of the free current carriers (n is the concentration of $Na^+O^-SiO_{3/2}$ structural elements); i.e., at about 1300°C only 2-2.5% of the Na^+O^- polar structural groups is dissociated and Si–O bonds are broken in only 0.001% of the structural elements.

It also follows from the foregoing that the following types of relaxation losses must be distinguished in alkali oxide glasses: a) those due to localized displacements of a large number of bound alkali cations within each dipolar structural element, and b) those due to displacement of a small number of free cations, determining steady conduction. For example, at $\varepsilon_i \approx 25$ kcal/mole, $\varepsilon_a \approx 5$ kcal/mole, and 500 K, at the instant when the external field is applied the number of bound cations displaced in 1 cm^3 of the glass is $n\nu \exp(-\varepsilon_a/kT)$ sec^{-1}; the average probability of displacement of all the cations is $f_a = \exp(-\varepsilon_a/kT) \approx 10^{-2}$. The number of moving free cations is $n_i\nu \exp(-\varepsilon_a/kT) = n\nu \exp[-(0.5\,\varepsilon_i + \varepsilon_a)/kT]$; alternatively, the probability of displacement (average value for all the cations in the glass) is $f_b = \exp[-(0.5\,\varepsilon_i + \varepsilon_a)/kT] \approx 10^{-8}$. At the initial instant of application of the external field the losses due to limited displacements of a considerable number of cations can be 10^6 times as great as the losses due to displacements of a small number of free cations.

If the alkali oxide content of the glass is low, cation displacements limited by the size of the microdisperse associated polar inclusions in the main polar medium are also possible [15].

The correlation of experimental data on the conductivities of numerous glasses investigated by various workers, compiled by L. A. Grechanik, E. A. Fainberg, and I. N. Zertsalova [16], is of considerable interest. The resultant graph shows that there is a regular relationship between the logarithm of the conductivity, log σ, and the energy of conduction ε_σ. In particular, the relation observed at 200°C can be, in the first approximation, represented as:

$$\log \sigma_{\text{exptl}} \approx 2.4 - 5.6\varepsilon_\sigma \tag{1}$$

where σ is in $ohm^{-1} \cdot cm^{-1}$ and ε_σ is in eV.

The statistical preexponential factor in the conductivity expression

$$\sigma = \sigma_0 \exp\left(-\frac{\varepsilon_\sigma}{2kT}\right) \tag{2}$$

can be calculated theoretically. Both for ionically conducting glasses and for semiconductor glasses the conductivity modulus is approximately constant:

$$\log \frac{\sigma_0}{n} \approx \log \frac{\sigma_0}{[v]} \approx 4 \tag{3}$$

where n and [v] are the concentration of polar structural elements and covalent bonds, respectively, in $moles/cm^3$. For the glasses studied by L. A. Grechanik, E. A. Fainberg, and I. N. Zertsalova the values of n and [v] are close in order of magnitude to 10^{-2} $mole/cm^3$. As the numerical value of 2 kT at 200°C (473K) is $8.15 \cdot 10^{-2}$ eV, we have, after substitution of the approximate value $\log \sigma_0 \approx 2$, derived from (3), into Eq. (2),

$$\log \sigma_{\text{theor}} \approx 2 - \frac{\varepsilon_\sigma}{2.3 \cdot 8.15 \cdot 10^{-2}} = 2 - 5.3\varepsilon_\sigma \tag{4}$$

It is seen that the theoretical expression (4) for log σ as a function of ε_σ at 200°C and $n \approx [v] \approx 10^{-2}$ is in satisfactory agreement with the empirical expression (1).

In conclusion, it should also be noted that it was recently found that electronically conducting glasses containing polar groups conform to the empirical relationship between the conduction energy ε_σ and the concentration n of dipolar structural units

$$\varepsilon_\sigma n^{1/4} = \text{const} \tag{5}$$

found earlier for silicate glasses with ionic conduction [17].

Literature Cited

1. R. L. Myuller, Collection: The Glassy State, Izd. Akad. Nauk SSSR (1960) pp. 61 and 111. [English translation: The Structure of Glass, Vol. 2, Consultants Bureau, New York (1960) pp. 50 and 215.]
2. W. M. Latimer, The Oxidation States of the Elements and their Potentials in Aqueous Solution, 2nd edn., New York (1952).
3. F. D. Rossini, D. D. Wagman, W. H. Evans, S. Levin, and V. Jaffe, Selected Values of Chemical Thermodynamic Properties, Nat. Bur. Std., Washington (1952).
4. V. I. Vedeneev, L. V. Gurvich, V. N. Kondrat'ev, V. A. Medvedev, and E. L. Frankevich, Chemical Bond Energies, Ionization Potentials, and Electron Affinity, Izd. Akad. Nauk SSSR (1962).
5. R. L. Myuller, L. A. Baidakov, and Z. U. Borisova, Vestn. Leningr. Gos. Univ. (22):77 (1962).
6. R. L. Myuller, L. A. Baidakov, and Z. U. Borisova, Vestn. Leningr. Gos. Univ. (10):94 (1962).
7. L. A. Baidakov, Z. U. Borisova, and R. L. Myuller, Zh. Prikl. Khim., 34:2446 (1961).
8. T. S. Moss, Photoconductivity in the Elements, London (1952).
9. R. L. Myuller, Zh. Prikl. Khim., 35:541 (1962).
10. Z. U. Borisova, R. L. Myuller, and Chin Cheng-ts'ai, Zh. Prikl. Khim., 35:774 (1962).
11. E. Mooser and W. B. Pearson, Nature, 190:406 (1961).
12. R. L. Myuller, Zh. Tekhn. Fiz., 25:1567 (1955).
13. R. L. Myuller, Zh. Tekhn. Fiz., 25:1556 (1955).
14. A. F. Borisov and V. I. Zadumin, this collection, p. 97.
15. R. L. Myuller, Zh. Tekhn. Fiz., 26:2614 (1956).
16. I. N. Zertsalova, L. A. Grechanik, and E. A. Fainberg, this collection, p. 74.
17. A. V. Danilov and R. L. Myuller, Zh. Prikl. Khim., 35:2012 (1962).

COMBINED STUDY OF CERTAIN ELECTRICAL AND MECHANICAL PROPERTIES OF GLASS AS A MEANS FOR REVEALING ITS STRUCTURAL PECULIARITIES

O. V. Mazurin

Combined study of a group of glass properties unified by a single mechanism often proves very fruitful. One such group can be defined among the electrical properties of glasses with ionic conductivity. These are the properties associated with activated displacement of charge carriers in the glass. We propose to describe them as migrational properties. They include electrical conductivity, diffusion of charge carriers, migrational dielectric losses (the sum of conduction and relaxation losses), changes of dielectric constant associated with relaxation processes (the difference $\varepsilon_0 - \varepsilon_\infty$), decrease of current due to thermal ionic polarization, and finally, certain regions of the temperature – frequency relationships of internal friction. Although this last is a mechanical property, it should be considered together with electrical properties.

This article is a discussion of the relations between individual migrational properties and of the use of data on these properties for elucidation of certain features of glass structure. Only glasses with alkali conduction, which have been studied most fully, are considered.

Of all the migrational properties of glasses, electrical conductivity is measured most easily and reliably, and therefore it has been studied in most detail. Electrical conductivity should therefore be regarded as the fundamental, "basic" migrational property.

K. K. Evstrop'ev's extensive investigations [1] have shown that there is a close connection between the conductivity of glass and the diffusion coefficients of the charge carriers. In glasses containing alkali-metal ions of the same size the diffusion coefficient can be calculated from the conductivity with an error not exceeding the aggregate error in the determinations of the respective values.

Recently it was shown by Isard [2] and (independently) by the present author [3] that a clear relation exists also between conductivity and migrational losses in glasses. We put forward the following empirical formulas (Isard gave only a graphical relation, less convenient to use):

$$\begin{aligned} &\text{for } (\log f + \log \rho) > 11 \quad \log \varepsilon'' = -1.9 + 40 \cdot 10^{-0.1(\log f + \log \rho)} \\ &\text{for } (\log f + \log \rho) \leqslant 11 \quad \log \varepsilon'' = 12.25 - (\log f + \log \rho) \end{aligned} \qquad (1)$$

where f is the frequency of the alternating field, ρ is the resistivity of the glass, and ε'' is the loss factor.

The formulas (1) proved valid (with an error not exceeding 0.3 of an order of magnitude) for various silicate and borosilicate glasses investigated both by us and by others and containing from 8 to 40% alkalies, from 0 to 40% bivalent oxides, etc. Thus it is possible even now to calculate approximately the temperature – frequency relationships of the loss factor from conductivity data.

Isard [2] also found a correlation between conductivity and the relaxation component of the dielectric constant.

Finally, it is known that the current decrease due to thermal ionic polarization and dielectric relaxation losses are based on the same mechanism. The dependence of the current decrease on time as the result of thermal ionic polarization can be calculated with a high degree of accuracy from the frequency – loss factor

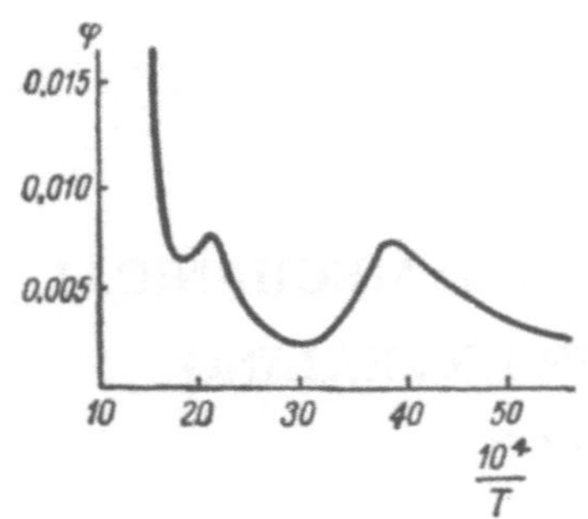

Fig. 1. Variation of the internal friction with the reciprocal absolute temperature for a chilled glass specimen of the composition 19% Na_2O, 81% SiO_2 at 0.78 cps [5].

relationship for a dielectric [3]. Thus, the dependence of nonsteady conductivity on time can also be expressed as a function of the specific conductance of glass.

It is therefore possible to devise a system of relationships between a large group of electrical properties and diffusion. Such relationships are as yet approximate. It is quite probable that they are not universal and should be modified somewhat when the glass composition is changed considerably. All this remains to be found. However, further refinement of the relations between these glass properties appears to us to be very promising.

It may be noted that variations of the resistance of glasses during polarization near an electrode can also be quantitatively correlated with their conductivity [4]. It is also known that the breakdown voltage for thermal breakdown is estimated from the conductivity and dielectric losses.

We now turn to internal friction. The low-temperature maxima on the plots of this property as a function of temperature (Fig. 1) are usually associated with displacement of charge carriers. This is proved by comparison of the effective activation energies calculated from conductivity and internal friction; they are usually found to have similar values. However, most determinations of internal friction are at present carried out in a narrow frequency range; this leads to large errors in determinations of activation energies. Therefore we consider it more reliable to compare the absolute conductivity values and the position of the frequency maximum of the coefficient of internal friction at the same temperature.

We now find the relation between $\varkappa_T$ and f_T, where f_T is the position of the frequency maximum of the coefficient of internal friction at temperature T, and $\varkappa_T$ is the conductivity at the same temperature.

We know that

$$f_T = \nu e^{-u_0/kT} \tag{2}$$

where ν is the thermal vibration frequency, u_0 is the effective activation energy (sum of half the dissociation energy u_d and the activation energy u_a), and k is the Boltzmann constant.

We consider that certain corrections must be applied to the Frenkel' equation for conductivity:

$$\varkappa_T = \frac{n_0 q^2 \delta^2 \nu \gamma}{3kT} e^{-u_0/kT} \tag{3}$$

where n_0 is the number of current-carrying ions in 1 cm^3 of glass, q is the ionic charge, and δ is the distance of a single ionic transfer.

The value of γ is associated with steric hindrance to transfer of alkali-metal ions. The Frenkel' theory assumes equal possibility of transfer of a dissociated ion in any direction. In reality, u_0 characterizes the required activation energy only for the impact of an ion into a space between the surrounding oxygen ions. Therefore γ is the ratio of the area of "free" regions of the potential sphere of the dissociated ion, not occupied by oxygen ions, to the total area of that sphere.

At present, only a very rough estimate of γ is possible. It may be concluded from a number of considerations that γ is between 0.3 and 0.1. Stevels [6] erroneously attributes values greater than unity to an analogous quantity.

The value of δ is apparently close to 5 A (the round sum of the diameters of sodium and oxygen ions). We take it to be from 4 to 8 A.

TABLE 1. Experimental Values of $\log(f_T/\varkappa_T T)$ for Different Glasses

Glass composition, mole %	T, K	$\log \frac{f_T}{\varkappa_T T}$	Literature source	
			f_T	$\varkappa_T$
19 Na_2O, 81 SiO_2	265	8.4	[5]	[3] *
15 Na_2O, 85 SiO_2	273	8.9	[7]	[3] *
30 Na_2O, 70 SiO_2	238	8.8	[7]	[3] *
33 K_2O, 67 SiO_2	252	8.5	[7]	[3] *
24 Na_2O, 12 CaO, 64 SiO_2	272	8.9	[8]	[3] *
24 Na_2O, 12 BaO, 64 SiO_2	293	8.9	[8]	[3] *

*Extrapolation.

Combining Eqs. (2) and (3) and assuming that the numerical values of ν and u_0 are the same in both (which should be the case if both equations represent the same process), we obtain

$$\frac{\varkappa_T T}{f_T} = \frac{n_0 q^2 \delta^2 \gamma}{3k} \tag{4}$$

For the glasses studied, containing from 13 to 33% R_2O, n_0 varies little. We take the value corresponding to a glass with 20% Na_2O, $1 \cdot 10^{22}$ cm^{-3}, for n_0. Substituting the above ranges of γ and δ values, we have:

$$\log \frac{f_T}{\varkappa_T T} = 7.9 - 9$$

This result is in good agreement with experimental data (Table 1).

It must be pointed out that polyalkali glasses deviate considerably from this relationship; this matter requires special study.

We now consider what information on the structure of glass can be obtained by studies of the dependence of their migrational properties on temperature and frequency.

We believe that a considerable proportion of the problems relating to glass structure reduces to the problem of microheterogeneity.

In our view, all the possible types of microheterogeneity in glasses are suitably subdivided into three groups: fluctuational, chemical, and physical.

Fluctuational heterogeneities, determined by random deviations of bond angles, interionic distances, and ionic positions, are apparently the most characteristic of any glassy state. These heterogeneities are grouped around a certain statistical mean structure. The greater the deviations from this mean structure, the rarer they are. The energy characteristics of any types of bond in glasses are correspondingly subject to fluctuational uncertainty.

Chemically heterogeneous glasses are taken to be glasses containing two or more types of regions with statistical mean structures differing in composition. The transition between such regions is evidently diffuse and does not correspond to a phase boundary. Nevertheless, the main bulk of each kind of region must have a quite definite statistical mean composition.

By studying the temperature dependence of internal friction it is possible in principle to obtain detailed information on the amounts of different alkali-containing structures in a glass. Unfortunately, however, the nature of the high-temperature maximum is still obscure [9]. Conclusive interpretation of the internal friction spectrum is therefore impossible at this time.

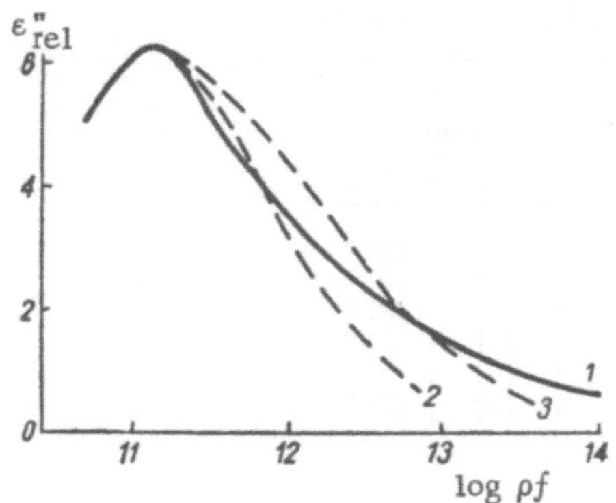

Fig. 2. Comparison of the experimental relaxation curve for alkali glasses (1) with the curves calculated from the Wagner equation for b = 0.6 (2) and b = 0.35 (3).

Let us now examine the conclusions which may be drawn from data on the temperature – frequency relationships of dielectric losses.

As the functions (1) are general for different glasses and temperatures, it follows that the relaxation component for different glasses is represented by similar curves. The general form of this relation is shown in Fig. 2. The same figure shows the course of two curves characterizing relaxation maxima in accordance with Wagner's formula for relaxation time distribution [10]. It is clear from the figure that for the experimental curve the distribution density b of the relaxation times does not remain constant but increases with the absolute value of log τ/τ_0 (i.e., with increasing deviation of log ρf from the value 11.2, corresponding to the maximum ε''_{rel}).

It is very important that, according to recent results [11], the temperature dependence of conductivity and of average relaxation time of dielectric relaxation losses are the same. The applicability of expressions (1) to glasses at a great variety of temperatures is further evidence of this. The mechanism of dielectric relaxation loss therefore includes both dissociation of the alkali-metal ion and its activated transfers.

There are two possible causes of such a relaxation process – fluctuational scatter of activation energies, and the existence of regions with especially high energy barriers in the glass.

Fluctuational scatter of activation energies should make the relaxation loss maximum more diffuse with decreasing temperature or vibration frequency; this has never yet been observed. More careful determinations are necessary in this field. If it is confirmed that temperature and frequency have no effect on the sharpness of the relaxation maximum, this would provide weighty evidence for the existence of nonconducting or weakly conducting inclusions in alkali glasses, most probably of the chemical heterogeneity type. These inclusions should apparently be regarded as nonpolar or weakly polar regions in a polar medium. The fraction of the glass volume occupied by these nonpolar structures in multialkali glasses probably alters relatively little with the composition of the glass.

At the same time, the broad distribution of relaxation times characteristic of glasses cannot be completely explained by the mechanism just described. It is most likely that in the region where log τ/τ_0 is greater than 3 or 4 (i.e., far from the main relaxation maximum – see Fig. 2) the dielectric losses are determined by activated displacements of undissociated ions by the mechanism postulated by R. L. Myuller [12].

TABLE 2. Comparison of the Effective Activation Energies of Certain Compounds in the Crystalline and Glassy States [3]

Formula of compound	u_0, eV	
	Polycrystalline	Glassy
$Na_2O \cdot SiO_2$	1.06	0.48
$Na_2O \cdot 2SiO_2$	1.40	0.55
$Li_2O \cdot 2SiO_2$	1.34	0.68
$Na_2O \cdot 2CaO \cdot 3SiO_2$	1.40	1.04
$Na_2O \cdot 3CaO \cdot 6SiO_2$	1.37	1.15
$2Na_2O \cdot CaO \cdot 3SiO_2$	1.25	0.77

According to the dielectric loss theory, the existence of enclosed conducting inclusions in a nonconducting medium should give rise to characteristic maxima on the temperature – frequency relationships of the dielectric losses. The discovery of such maxima for low-alkali sodium silicate glasses [13] is convincing evidence in favor of R. L. Myuller's views on the structure of these glasses [12]. In our opinion, study of the temperature – frequency relationships of losses is the most reliable modern method not only for detecting but also for studying (from the value of the effective activation energy and the statistical factor) enclosed polar inclusions in a nonpolar glassy medium.

We now turn to physical heterogeneities. Experimental data show that the effective activation energy of alkali compounds is considerably less in the glassy than in the crystalline state (Table 2). The difference is especially large for the simplest compounds.

The data in the table demonstrate the enormous influence of small changes in the cell structure of the silicon–oxygen lattice on the strength of bonding of alkali-metal ions in it. Apparently, purely steric factors have much more influence on the mobility of alkali-metal ions in glass and crystals than is usually believed.

The fact of greatest importance for us, however, is the great decrease in the mobility of alkali-metal ions in the transition from a deformed to a regular lattice. Even if crystal-like regions (crystallites) are present in glass, their presence should not affect the temperature – frequency relationships of the electrical properties of the glass.

The situation is different with regard to internal friction. Here we should probably find absorption bands corresponding to movement of alkali-metal ions in the crystalline regions. However, in order to detect crystallites by this method we must study the influence of glass crystallization on internal friction. Apart from the paper by Day and Rindone [14], there is no information on this subject in the literature. Unfortunately, Day and Rindone studied internal friction at only one frequency, which makes interpretation of their absorption bands very difficult.

Literature Cited

1. K. K. Evstrop'ev, Author's summary of candidate's dissertation, Leningr. Gos. Univ. im. A. A. Zhdanova (1962).
2. J. Isard, Proc. Inst. Elec. Engrs., London, 109B (Suppl. 22):440 (1962).
3. O. V Mazurin, "Electrical properties of glass," Tr. Leningr. Tekhnol. Inst. im. Lensoveta (62) (1962) [a translation of portions of this book has been included as an introduction to this collection]; O. V. Mazurin, Author's summary of doctoral dissertation, Leningr. Tekhnol. Inst. im. Lensoveta (1962).
4. V. B. Brailovskii, M. Ya. Rozenblyum, and O. V. Mazurin, this collection, p. 102.
5. I. Mohyuddin and R. Douglas, Phys. Chem. Glasses, 1:71 (1960).
6. J. Stevels, Electrical Properties of Glass [Russian translation], IL (1961).
7. H. Rötger, Glastech. Ber., 31:54 (1958).
8. R. Ryder and G. Rindone, J. Am. Ceram. Soc., 43:662 (1960).
9. R. W. Douglas, Phys. Chem. Glasses, 4:34 (1963).
10. K. Wagner, Ann. Phys., 40:817 (1913).
11. H. Taylor, Trans. Faraday Soc., 52:873 (1956); R. Charles, J. Am. Ceram. Soc., 45:105 (1962).
12. R. L. Myuller, Zh. Vses. Khim. Obshchestva im. D. I. Mendeleeva, 8:197 (1963).
13. V. K. Leko and M. L. Dorokhova, this collection, p. 118.
14. D. Day and G. Rindone, J. Am. Ceram. Soc., 44:161 (1961).

VARIATIONS OF ACTIVATION ENERGY AND VOLUME CONDUCTIVITY OF SOLID GLASSES IN RELATION TO THE CONDUCTION MECHANISM

I. N. Zertsalova, E. A. Fainberg, and L. A. Grechanik

We discovered earlier [1] the existence of a definite relationship between E, the activation energy of conductivity of glasses, their volume resistivity ρ_T, and the conduction mechanism: glasses with predominantly electronic conduction have lower activation energies than glasses with predominantly ionic conduction having the same volume resistivity at the same temperature. It is demonstrated in the present paper that this relationship is valid for the great majority of glass systems.

We collected volume conductivity data on more than 1500 solid glasses. Most of the data were taken from the literature [1-32], and some of our own results were also used. The specific volume resistivity at 200°C (ρ_{200}), ρ_0 (at T = ∞), and the activation energy E of the current carriers were calculated for all the glasses. The following formula was used for all the calculations:

$$\rho_T = \rho_0 e^{\frac{E}{2kT}}$$

It must be noted that extrapolation of linear plots of log ρ versus 1/T gave results devoid of physical meaning for a number of glasses. For example, at 200°C most chalcogenide glasses either are in the liquid state or crystallize extensively. However, the validity of these assumptions for plotting the graphs is obvious.

Plots of log ρ_0 versus E and log ρ_{200} versus E are shown in Figs. 1 and 2, respectively. These graphs are based on 280 glass compositions chosen so as to cover the maximum number of systems, with each system adequately represented. Because of the large amount of tabular data, the chemical compositions and conductivity data are not given in this paper.

Let us examine the experimental data in more detail.

Glasses of the vanadium phosphate and tungsten phosphate groups [20] are electronic conductors. This can be taken as proved in [24, 25]. It should be noted that glasses with high R_2O contents in these systems (0.85; +0.49), (1.28; +0.40)* are also contained in region A.

Electronic conduction in glasses containing iron oxides is assumed in [1, 19]. Iron-containing glasses in which the proportion of electricity transferred by other carriers appears to be small are found in the field of glasses with electronic conduction on the diagrams. For example, if iron and sodium oxides are introduced simultaneously into glasses, the points move from the electronic to the cationic conduction field with increase of the $Na_2O:Fe_2O_3$ ratio [1].

All glasses with one or more alkali-metal ions are in field B.

* Here and subsequently, the first number in parentheses is the activation energy and the second is the value of log ρ_0.

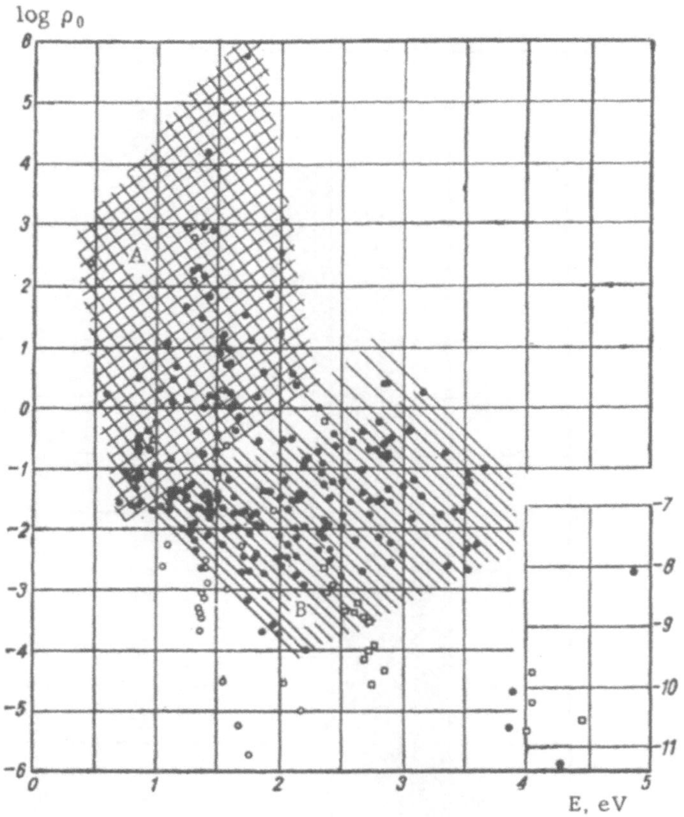

Fig. 1. Relation between activation energy and ρ_0 of glasses. A) Region of glasses having proven or presumed electronic conduction; B) region of glasses having proven or presumed cationic conduction: O) chalcogenide glasses; □) fluoride glasses; ●) other glasses.

Evstrop'ev and Khar'yuzov [9] showed convincingly that barium ions are the current carriers in binary barium silicate glasses. In fact, glasses in the system $BaO-SiO_2$ (2.90; −0.46), (2.78; −0.65), (2.67; −1.0) are found in region B in the diagram.

Lindner et al. [28] investigated diffusion of Pb ions into glass by a radioactive isotope method and found that the activation energies of self-diffusion and conduction are equal. This experiment provides weighty evidence in favor of current transfer by bivalent lead ions in lead glasses. In Fig. 1 glasses in the system $PbO-SiO_2$ (2.72; −1.52), (2.30; −0.69), (2.15; −1.40), (2.20; −1.95) are in field B. It may be noted that glasses in the system $PbO-Al_2O_3-SiO_2$ are in the same field.

Alkali-free silicate and borate glasses containing bivalent oxides are in the cationic conductor field.

Most authors [3, 5, 29] consider that the conduction of quartz glass is due to alkali-metal ions present as impurities. In fact, the points for glassy silica (2.72; −0.65), (2.34; −0.90) are in field B in the diagram.

The high activation energies of glasses in the system $CaO-B_2O_3-Al_2O_3$ prompted Owen [18] to postulate a new type of conduction in them (migration of oxygen ions). However, these glasses, e.g., (3.30; −1.75), (3.54; −1.23), (3.66; −0.98) do not deviate from the general relationship. This is particularly clear in Fig. 2 from the upper points of the cationic field.

Glassy GeO_2 (2.1; + 0.58) is in field A, but a glass of the composition (mole %): GeO_2 − 97.5 and PbO − 2.5 (3.0; −2.40) is in region B [11].

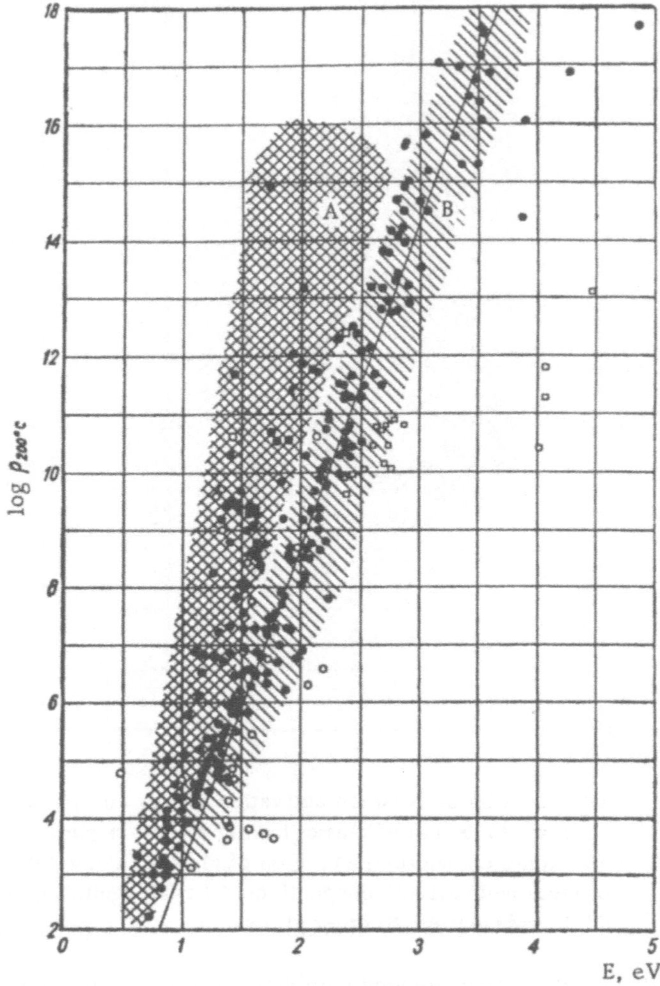

Fig. 2. Relation between activation energy and volume resistivity of glasses at 200°C. (The symbols in Fig. 2 are explained in the caption to Fig. 1.)

Glass of the composition (mole %): SiO_2 – 40, BaO – 25, TiO_2 – 35, melted in a neutral atmosphere, lies in field B (3.15; –1.44). The same glass, but melted in a reducing atmosphere, lies in field A (1.67; –0.13). It may be noted that certain authors [30, 31] correlate semiconductor properties with the presence of transition elements in different valence states in the glasses.

The group of fluoride glasses does not conform to the general relationship. This is possibly due to the anionic conduction presumed in such glasses. However, the literature data on conductivity of such glasses are quite inadequate for combining them in an independent region in the diagram [23].

Several workers have demonstrated electronic conduction in chalcogenide glasses. However, most such glasses known to us do not conform to the presumed relationship. R. L. Myuller [22] showed that the electrical resistance of chalcogenide glasses is greatly dependent on the cooling conditions. Chilled specimens are in the electronic region of the diagram, but the same glasses after annealing lie in the cationic field. It is possible that here (and in some other cases) the glasses undergo structural changes [32] or even crystallize. It is reasonable to suppose that they are excluded from the proposed relationship either by such effects or by the nature of the glasses themselves (oxygen-free glasses).

Summary

1. The relationship found between activation energy, volume resistivity, and conduction mechanism of glasses is valid for all the systems examined apart from fluoride and chalcogenide glasses.

2. At this time there are only two methods for direct determination of current carriers in glasses (investigation of ion diffusion by the radioactive isotope method, and Tubandt's technique), both of which are rather complicated experimentally. In view of this, and of the universal nature of the relationship reported in this paper, there is reason to believe that it may prove useful for determining the type of conduction in glasses.

Literature Cited

1. L. A. Grechanik, E. A. Fainberg, and I. N. Zertsalova, Fiz. Tverd. Tela, 4(2):454 (1962).
2. A. F. Ioffe, Izv. Petrogradsk. Politekhn. Inst. (1915).
3. A. Owen and R. Douglas, J. Soc. Glass Technol., 43(211):159T (1959).
4. S. A. Shchukarev and R. L. Myuller, Zh. Fiz. Khim., 1:625 (1930).
5. O. V. Mazurin, "Electrical properties of glass," Tr. Leningr. Tekhnol. Inst. im. Lensoveta (62) (1962). [A partial translation of this book appears as an introduction to this collection.]
6. O. V. Mazurin, Author's summary of candidate's dissertation, Leningr. Tekhnol. Inst. im. Lensoveta (1953).
7. B. I. Markin, Author's summary of candidate's dissertation, Leningr. Gos. Univ. im. A. A. Zhdanova (1951).
8. T. M. Makarova, O. V. Mazurin, and V. S. Molchanov, Izv. Vuzov, Khimiya i Khim. Tekhnol. (6):1072 (1960).
9. K. K. Evstrop'ev and V. A. Khar'yuzov, Dokl. Akad. Nauk SSSR, 136(1):140 (1961).
10. I. G. Mel'nikova, K. S. Evstrop'ev, and A. Ya. Kuznetsov, Zh. Fiz. Khim., 25(11):1318 (1951).
11. K. S. Evstrop'ev and A. O. Ivanov, Optiko-Mekh. Prom. (9):1 (1959).
12. L. A. Grechanik, E. A. Fainberg, and I. N. Zertsalova, Zh. Prikl. Khim., 36(1):91 (1963).
13. O. V. Mazurin and V. B. Brailovskii, Izv. Vuzov, Ser. Fiz. (1) (1959).
14. B. I. Markin, Fiz. Tverd. Tela, 3:450 (1961).
15. J. Isard, J. Soc. Glass Technol., 43(211):113T (1959).
16. E. A. Antonova, Author's summary of candidate's dissertation, Leningr. Tekhnol. Inst. im. Lensoveta (1954).
17. O. V. Mazurin and G. T. Petrovskii, Tr. Leningr. Tekhnol. Inst. im. Lensoveta, Sbornik Studencheskikh Rabot (1956) p. 51.
18. A. Owen, Phys. Chem. Glass, 2 (3):87 (1961).
19. O. V. Mazurin, G. A. Pavlova, E. Ya. Lev, and E. K. Leko, Zh. Tekhn. Fiz., 27 (12):2702 (1957).
20. L. A. Grechanik, N. V. Petrovykh, and V. G. Karpechenko, Fiz. Tverd. Tela, 2:2131 (1960).
21. B. T. Kolomiets and T. F. Nazarova, Collection: Solid State Physics, Vol. II, Izd. Akad. Nauk SSSR (1959) p. 22.
22. L. A. Baidakov, Z. U. Borisova, and R. L. Myuller, Zh. Prikl. Khim., 34(11):2446 (1961).
23. G. T. Petrovskii, E. K. Leko, and O. V. Mazurin, Optiko-Mekh. Prom. (2):18 (1961).
24. P. Baynton, H. Rawson, and J. Stanworth, J. Electrochem. Soc., 104 (4) (1957).
25. I. I. Kitaigorodskii, V. K. Frolov, and Kuo-Cheng, Steklo i Keram. (12):5 (1960).
26. A. Ya. Kuznetsov, Author's summary of doctoral dissertation, Gos. Optich. Inst. im. S. I. Vavilova (1956).
27. K. S. Evstrop'ev, A. Ya. Kuznetsov, and I. G. Mel'nikova, Zh. Tekhn. Fiz., 21:104 (1951).
28. R. Lindner, W. Hassenteufel, and Y. Kotera, Z. Physik. Chem. (Frankfurt), 23(5/6):408 (1960).
29. S. Strauss, D. Moore, W. Harrison, and L. Richards, J. Research Nat. Bur. Standards, 56:135 (1956).
30. M. Munakata and M. Imamoto, Bull. Electrotech. Lab. (Tokyo), 24(2):90 (1960).
31. V. A. Ioffe, I. B. Patrina, and I. S. Poberovskaya, Fiz. Tverd. Tela, 2:656 (1960).
32. V. A. Khar'yuzov and K. S. Evstrop'ev, Optiko-Mekh. Prom. (10):17 (1961).

II

INVESTIGATION OF GLASSES WITH IONIC CONDUCTION

ELECTRICAL CONDUCTIVITY OF MOLTEN GLASSES

K. A. Kostanyan, K. S. Saakyan, and É. M. Avetisyan

Investigation of the conductivity of molten alkali glasses also containing bivalent ions is of great theoretical and practical interest.

During the past few years we have investigated electrical conductivity of glasses in the systems

$$Na_2O - RO - SiO_2$$

where RO = CaO, MgO, PbO, ZnO, BaO, BeO, CdO, in the range of 900-1400°C. These investigations show that in the range of 1100-1400°C the conductivity of these glasses obeys the well-known exponential relation

$$\log \varkappa = A - \frac{B}{T} \tag{1}$$

where A and B are constants and T is the absolute temperature.

There is a proportionality relation between conductivity and the sodium oxide content for all the systems studied. In triangular diagrams of the systems

$$Na_2O - RO - SiO_2$$

the isotherms corresponding to compositions of equal conductivities are almost parallel to the $RO-SiO_2$ side; i.e., these isotherms correspond to compositions of constant sodium oxide content.

As R. L. Myuller points out [1], for a more rigorous evaluation of the influence of composition on the conductivity of alkali glasses it is necessary to use molar concentrations of the alkali ion per unit volume of the glass. This approach to assessment of the influence of composition on conductivity proved to be especially fruitful with molten glasses, where conductivity is determined primarily by the sodium content. Figure 1 shows the isotherm (for 1200°C) of the resistivities of glasses of different composition plotted against the sodium ion concentration.

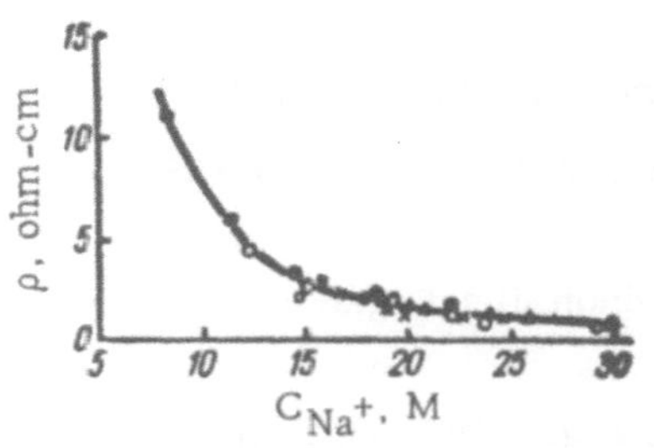

Fig. 1. Variations of resistivity of glasses in the system $Na_2O-RO-SiO_2$ with the alkali ion concentration at 1200°C. Δ) $Na_2O-ZnO-SiO_2$; ×) $Na_2O-CdO-SiO_2$; ○) Na_2O-SiO_2; ●) $Na_2O-CaO-SiO_2$; □) $Na_2O-PbO-SiO_2$.

It is seen in the diagram that all the points lie on the same curve, regardless of the nature and concentration of the bivalent cation.

The curve shows that the principal factor determining conductivity of these glasses in the molten state is the concentration of the alkali ions as the main current carriers.

It is clear that bivalent cations have very little influence on the mobility of sodium ions and play little part in the transfer of electricity. It must be pointed out that all this refers to

Glass No.	Sodium silicate glasses: $[M]\cdot 100$, moles/ml	λ, $ohm^{-1}\cdot cm^{-1}$	Ψ_{Φ}, kcal/mole	P_e	$\Delta H^{\ddagger}$ kcal/mole	$\Delta S^{\ddagger}$ cal/mole	$\Delta Z^{\ddagger}$ kcal/mole	Glass No.	Sodium borate glasses: $[M]\cdot 100$, moles/ml	λ $ohm^{-1}\cdot cm^{-1}$	Ψ_{Φ}, kcal/mole	P_e	$\Delta H^{\ddagger}$ kcal/mole	$\Delta S^{\ddagger}$ cal/mole	$\Delta Z^{\ddagger}$ kcal/mole
1	2.97	32.1	13.0	2.61	6.5	—5.86	14.0	1-Na	1.95	32.1	29.0	4.09	14.5	0.33	14.1
2	2.48	33.6	13.5	2.68	6.8	—5.65	14.0	1*a*-Na	1.77	23.5	31.6	4.13	15.8	0.74	14.9
3	2.24	24.1	16.25	2.77	8.1	—5.23	14.8	2-Na	1.61	20.1	33.0	4.17	16.5	0.97	15.3
4	1.94	16.4	20.50	2.96	10.2	—4.34	15.8	2*a*-Na	1.43	20.0	34.0	4.20	16.9	1.30	15.3
5	1.53	15.35	18.00	2.74	9.0	—5.42	15.95	4-Na	1.26	17.6	35.8	4.25	17.9	1.80	16.5
6	1.245	11.1	19.9	1.75	9.9	—5.35	16.8	5-Na	0.96	13.5	38.4	4.39	19.2	2.30	16.3

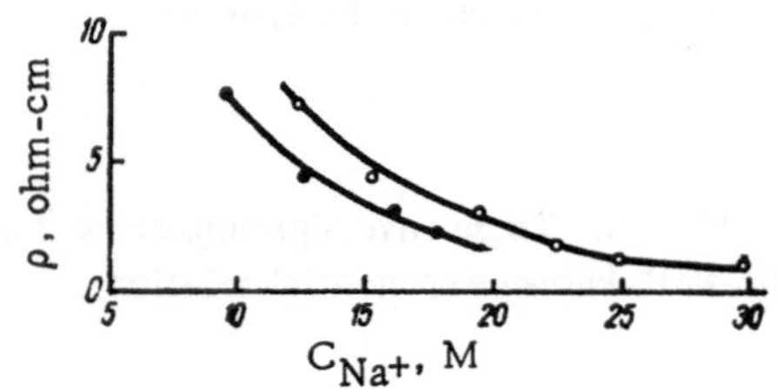

Fig. 2. Variations of resistivity of sodium silicate and sodium borate glasses with the alkali ion concentration at 1000°C. O) Na_2O-SiO_2; ●) $Na_2O-B_2O_3$.

glasses with high sodium concentrations ($Na_2O > 10$ M) and relatively low contents of bivalent cations.

At lower sodium concentrations the relationship is of a more complex character.

In discussing the influence of bivalent modifier ions on the conductivity of molten glasses, we must also consider the influence of the glass-forming cation. It is of great interest in this connection to compare the conductivities of molten sodium silicate and sodium borate glasses.

By R. L. Myuller's theory of glass conductivity [1], the constant A in Eq. (1) is connected with the alkali ion concentration M by the expression

$$P_e = A - \log[M] \tag{2}$$

while the relation between B and the dissociation energy Ψ_{Φ} is given by the expression

$$\Psi_{\Phi} = 4.6RB \tag{3}$$

According to R. L. Myuller, the energy Ψ_{Φ} for molten sodium silicate glasses is 22-26 kcal/mole, while the mobility factor P_e is independent of the chemical composition of the glass and retains the same value (the theoretical value is 3.7 ± 1) even when the glass is melted.

The table gives values of P_e and Ψ_{Φ} for sodium silicate and sodium borate glasses at 1000°C [2, 3].

The table shows that $P_e < 3$ and $\Psi_{\Phi} < 20.5$ for sodium silicate glasses, and $P_e > 4$ and $\Psi_{\Phi} > 29.0$ for sodium borate glasses.

This difference between the values of Ψ_{Φ} for sodium borate and sodium silicate glasses is somewhat at variance with the conductivities of these glasses.

In Fig. 2 the resistivities of sodium silicate and sodium borate glasses at 1000°C are plotted against the alkali ion concentration. It is seen that at a given sodium ion concentration sodium borate glasses have higher conductivity than sodium silicate glasses. This is not accidental. Comparison of conductivity data for other alkali silicate and alkali borate glasses shows that they conform to the same law. At the same time, alkali silicate glasses have low values of Ψ_{Φ} and P_e, while the corresponding values for alkali borate glasses are relatively high.

This discrepancy between the values of Ψ_{Φ} and conductivities of alkali borates and silicates and the anomalous values of P_e for molten glasses are attributable to the entropy term included in constant A of Eq. (1).

The appearance of an entropy term in Eq. (1) when glass passes into the molten state was noted by R. L. Myuller [4]. On the basis of the transition state theory, Bockris et al.. derived an equation for the equivalent conductance of molten glasses, including an entropy term [5]. The equation is of the form

$$\lambda = 3.62 \cdot 10^{19} z d^2 \exp\left(-\frac{\Delta H^{\ddagger}}{RT}\right) \exp\left(\frac{\Delta S^{\ddagger}}{R}\right) \tag{4}$$

where z is the cation charge, d is half the distance between neighboring structural units, ΔH is the heat of activation, found from B in Eq. (1) with the aid of the expression

$$\Delta H^{\ddagger} = \frac{RB}{0.434} \tag{5}$$

and $\Delta S^{\ddagger}$ is the entropy.

The table gives values of $\Delta H^{\ddagger}$, $\Delta S^{\ddagger}$, and $\Delta Z^{\ddagger}$ (free energy of activation) for sodium silicate and sodium borate glasses at 1000°C. $\Delta Z^{\ddagger}$ is calculated from the expression

$$\Delta Z^{\ddagger} = \Delta H^{\ddagger} - T\Delta S^{\ddagger} \tag{6}$$

The values found for the free energy of activation for sodium silicate and sodium borate glasses provide an explanation of the higher conductivity of sodium borate glasses – at a given sodium ion concentration $\Delta Z^{\ddagger}$ is lower for borate than for silicate glasses (for example, compare glass 4 with glass 1-Na). This indicates that the conductivities of molten glasses are determined by changes of the free energy of activation $\Delta Z^{\ddagger}$.

The values of $\Delta Z^{\ddagger}$ are lower for sodium borate than for sodium silicate glasses because the values of $\Delta S^{\ddagger}$ for sodium borate glasses are positive, as shown by Eq. (4).

Increase of entropy in the conduction process is also characteristic for other alkali borate glasses. Molten alkali silicate glasses have negative values of $\Delta S^{\ddagger}$.

The different values of $\Delta S^{\ddagger}$ for alkali borate and alkali silicate glasses account for the high value of P_e for molten alkali borates and the low value of P_e for alkali silicates (see the table).

These entropy values, of opposite sign for alkali borate and alkali silicate glasses, are evidence of structural differences between borate and silicate glasses, due to the nature of the glass-forming cation.

Literature Cited

1. R. L. Myuller, Fiz. Tverd. Tela, 2:1333 (1960).
2. K. A. Kostanyan, Izv. Akad. Nauk Arm. SSR, Ser. Khim. Nauk XI:65 (1958).
3. K. A. Kostanyan and K. S. Saakyan, Izv. Akad. Nauk Arm. SSR, Ser. Khim. Nauk XIV:409 (1961).
4. R. L. Myuller, Zh. Tekhn. Fiz., 25:276 (1955).
5. J. O. M. Bockris, J. A. Kitchener, S. Ignatowicz, and J. M. Tomlinson, Trans. Faraday Soc., 48:75 (1952).

ELECTRICAL CONDUCTIVITY OF ALKALI SILICATE AND BOROSILICATE GLASSES CONTAINING OXIDES OF GALLIUM AND IRON

A. O. Ivanov and E. I. Galant

The following series of glasses were investigated:

$$13Na_2O\,(87-x)\,SiO_2\cdot xGa_2O_3$$
$$13Na_2O\,(87-x)\,SiO_2\cdot xFe_2O_3$$
$$13Na_2O\cdot 17B_2O_3\,(70-x)\,SiO_2\cdot xGa_2O_3$$

the coefficients in the formulas representing molecular percentages. In the first and second series SiO_2 was partially replaced by Ga_2O_3 at a constant Na_2O content, and in the third series the Na_2O and B_2O_3 contents were kept constant while SiO_2 was replaced by Ga_2O_3. The conductivity was measured in the range of 150-500°C. Figure 1 shows the effects of gallium and iron oxides on the resistivity of sodium silicate glasses; the resistivity isotherm of the aluminosilicate series [1] is shown for comparison. It is clear from this comparison that the effects of Ga_2O_3, Al_2O_3, and Fe_2O_3 (up to 13%) on the resistance of sodium silicate glasses are completely analogous. The first additions of these oxides raise the resistance; in the region up to 13% Me_2O_3 the resistance falls considerably; at Ga_2O_3 and Al_2O_3 concentrations above 13% the resistance rises again. In the case of Ga_2O_3 and Al_2O_3 the resistance is at a minimum when the ratio $Me_2O_3/Na_2O = 1$. Glasses containing gallium and iron have higher resistance than glasses with aluminum. This is due to the decrease of the polarizing power of the Me^{3+} cation with increasing radius, which intensifies interaction between oxygen and Na^+ ions and results in firmer bonding. The position of the resistance maximum also depends on the radius of the Me^{3+} cation: it lies at 2% Al_2O_3, 3.5% Ga_2O_3, and 5% Fe_2O_3.

The influence of Ga_2O_3 on the resistance of sodium borosilicate glasses is shown in Fig. 2. In this series replacement of SiO_2 by Ga_2O_3 results in a slight decrease of resistance at Ga_2O_3 concentrations from 0 to 9%. The resistance increases in the region from 9 to 25% Ga_2O_3. The dependence of the dissociation energy on composition for this series of glasses is of a nature similar to the resistivity isotherms.

In these series of glasses the volume concentration of Na^+ ions per cm^3 of glass remains almost unchanged and equal to the concentration in the original compositions. The changes of conductivity observed when SiO_2 is replaced by Ga_2O_3 and Fe_2O_3 are therefore due exclusively to changes in bonding of Na^+ ions and the corresponding changes in the dissociation energy. In the ranges from 2 to 13% Al_2O_3, from 3 to 13% Ga_2O_3, and from 5 to 13% Fe_2O_3 these oxides act as glass-formers. They can use the oxygen introduced with the alkali to build their own oxygen tetrahedrons, forming a single aluminum–, gallium–, or iron–silicon–oxygen network. The Na^+ ions pass from the silicate component into the interstices of the aluminate, gallium, or iron component and are held by forces of electrostatic interaction with the centers of the negatively charged tetrahedrons. The probable explanation of the fall of resistance in this region is that the gallium–, iron–, and aluminum–oxygen tetrahedrons are larger and therefore provide more freedom of movement for Na^+ ions. At the ratio $Me_2O_3/Na_2O = 1$. when all the oxygen from the alkali has been consumed, a part of the oxides Ga_2O_3 and Al_2O_3 (above 13%) enters the glass composition as modifiers (Me^{3+} cations in sixfold coordination); the resistance increases. We attribute the increase of resistance on introduction of small amounts of Ga_2O_3, Fe_2O_3, and Al_2O_3 to the fact that these oxides act as modifiers in the glass at low concentrations (Me^{3+} cations in sixfold coordination). The point

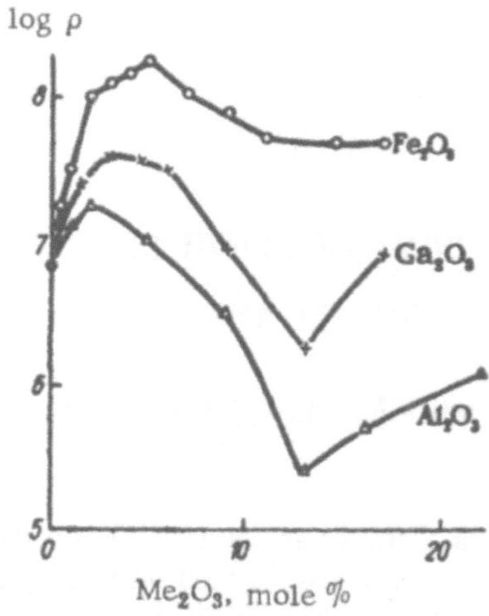

Fig. 1. Effect of replacement of SiO_2 by the oxides of gallium, iron, and aluminum on the resistivity of glasses in the series $13Na_2O \cdot (87 - x)SiO_2 \cdot xMe_2O_3$ at 150°C.

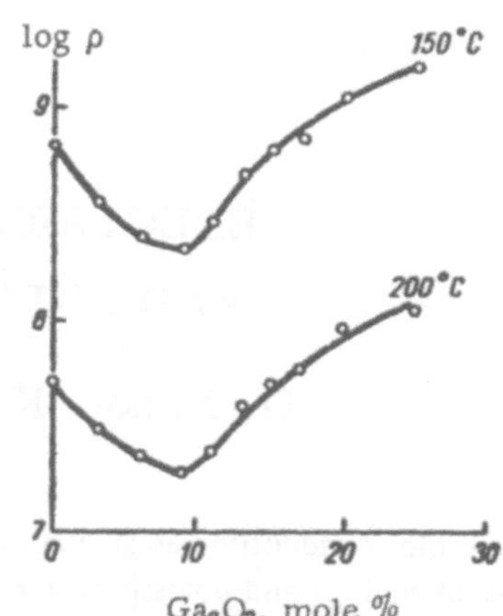

Fig. 2. Effect of replacement of SiO_2 by gallium oxide on the resistivity of glasses in the series $13Na_2O \cdot 17B_2O_3 \cdot (70 - x) \cdot xGa_2O_3$.

at which these oxides begin to take part in network formation depends on the ionic radius and hence the field strength of the Me^{3+} cation. The analogous effects of Fe_2O_3 and Al_2O_3 on the conductivity of alkali silicate glasses (in contrast to alkali-free glasses) and data on the diffusion of Na^+ in iron-containing glasses indicate absence of electronic conduction in this series (in the range from 0 to 13% Fe_2O_3).

In the series $13Na_2O \cdot 17B_2O_3 \cdot (70 - x)SiO_2 \cdot xGa_2O_3$, when SiO_2 is replaced by Ga_2O_3 the structural state of Ga_2O_3 and B_2O_3 is determined by the ratio $(Na_2O - Ga_2O_3)/B_2O_3$. The observed change of conductivity on introduction of Ga_2O_3 is the consequence of several factors: the influence of Ga_2O_3 itself, which forms gallium–oxygen tetrahedrons, simultaneous transition of boron from fourfold to threefold coordination, and transfer of sodium from the borate to the gallate component. The resistance minimum at 9% Ga_2O_3 is probably associated with the end of transition of boron into threefold coordination. The absence of an extremal at 13% Ga_2O_3 is probably due to the fact that gallium can exist in fourfold coordination in borosilicate glasses when $B_2O_3 \geq Ga_2O_3 \geq Na_2O$. The boron–oxygen trigonal units can act as oxygen donors in this case [2].

Literature Cited

1. O. V. Mazurin, "Electrical properties of glass," Tr. Leningr. Tekhnol. Inst. im. Lensoveta, (62) (1962) p. 69 [see this collection, p. 5.]
2. E. I. Galant, Dokl. Akad. Nauk SSSR, 141 (2):417 (1961).

ELECTRICAL CONDUCTIVITY OF SODIUM AND POTASSIUM GERMANATE GLASSES

A. O. Ivanov, K. S. Evstrop'ev, and M. L. Dorokhova

The molar conductivities of sodium and potassium germanate glasses are plotted against the volume concentrations of sodium and potassium in Fig. 1. Molar conductivity isotherms for binary alkali glasses of the silicate and borate systems are shown for comparison in Fig. 2. It is clear from these figures that the conductivity isotherms of sodium and potassium germanate glasses are completely similar to the isotherms of the corresponding borate systems. The isotherms have minima at the following concentrations: [Na] = $0.6 \cdot 10^{-2}$ g-at/ml and [K] = $0.8 \cdot 10^{-2}$ g-at/ml for germanate glasses and at [Me] = $0.5 \cdot 10^{-2}$ g-at/ml for borate glasses. In germanate and borate systems, from a certain alkali concentration, potash glasses have higher conductivities and lower activation energies than soda glasses. The variations of conductivity with composition of silicate glasses differ considerably from the corresponding variations for germanate and borate systems. Introduction of small amounts of alkali into quartz glass raises the conductivity snarply.

In our opinion, the conductivity variations with composition in the region of low Me_2O contents in silicate glasses are not of the same character as for borate and germanate glasses because of the different effects of alkalies on the structure of these glasses. When an alkali is introduced into a silicate glass, the alkali ions tear the oxygen–silicon network, loosen the structure, and weaken the framework. For this reason the first portions of Me_2O produce a considerable increase of conductivity. In borate and, according to our hypothesis [7], in

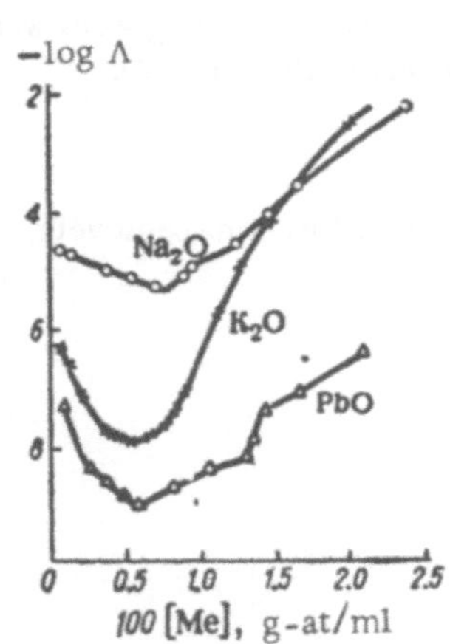

Fig. 1. Variations of the molar conductivity of germanate glasses with the volume concentrations of Na, K, and Pb ions in the systems Na_2O-GeO_2, K_2O-GeO_2, and $PbO-GeO_2$ at 300°C.

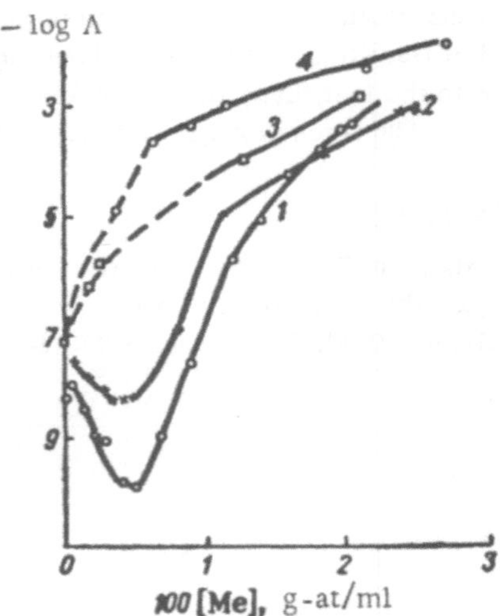

Fig. 2. Variations of the molar conductivity of two-component glasses with composition. 1) System $K_2O-B_2O_3$ [2]; 2) system $Na_2O-B_2O_3$ [3]; 3) system K_2O-SiO_2 [4]; 4) system Na_2O-SiO_2 [5, 6].

germanate glasses introduction of alkali does not cause rupture of bridge bonds. On the contrary, owing to transition of a part of the boron into fourfold coordination and of germanium into sixfold coordination, additional bonds are formed, and the strength and packing density of borate and germanate glasses increase, influencing a number of properties. Alkali ions in borate and germanate glasses are situated in interstices and are held by forces of electrostatic interaction with the centers of the negatively charged structural units. This accounts for the decrease of conductivity and a certain increase of activation energy when alkalies are introduced into these glasses. The conductivity begins to increase when contact between polar groups becomes possible.

The fact that at considerable alkali concentrations potassium germanate and borate glasses have higher conductivity than the corresponding sodium glasses is explained as follows. Potassium ions are held less firmly in the network than sodium ions. However, at low Me_2O concentrations, when the glasses have a compact and dense structure, the ion size becomes the predominant factor. Potassium ions are much less free to move in a compact network than sodium ions. At considerable Me_2O concentrations – above 20% for germanate glasses and above 30% for borate glasses – when the network is sufficiently loose and potassium can move as freely as sodium, the strength with which the ions are held in the network becomes the predominant factor and potassium glasses have higher conductivity.

Literature Cited

1. K. S. Evstrop'ev and A. O. Ivanov, Optiko-Mekh. Prom. (9) (1959).
2. B. I. Markin and R. L. Myuller, Zh. Fiz. Khim., 5 (9) (1934).
3. S. A. Shchukarev and R. L. Myuller, Zh. Fiz. Khim., 1 (6) (1930).
4. A. Ya. Kuznetsov and I. G. Mel'nikova, Zh. Fiz. Khim., 24 (10) (1950).
5. Seddon, J. Soc. Glass Technol., 16:450 (1932).
6. O. V. Mazurin, "Electrical properties of glass," Tr. Leningr. Tekhnol. Inst. im. Lensoveta (62) (1962). [A partial translation of this book is included as an introduction to this collection.]
7. A. O. Ivanov and K. S. Evstrop'ev, Dokl. Akad. Nauk SSSR, 145 (4):797 (1962).

ELECTRICAL CONDUCTIVITY OF CERTAIN FLUORIDE GLASSES

G. T. Petrovskii, E. K. Leko, and O. V. Mazurin

Although it has been known for over 30 years that glasses can be obtained from beryllium fluoride and fluorides of other metals, the properties of such glasses have been studied very little. The main characteristics studied were the optical properties, density, and tendency to glass formation [1-5]. The electrical properties have not been studied at all. We can refer only to the work of Izumitani and Terai [6] who investigated various properties of oxide–fluoride glasses in the systems $B_2O_3-BaF_2$ and $B_2O_3-PbF_2$ and noted that the conductivity of the glasses increases with increasing fluoride content. Nevertheless, study of the conductivity of pure fluoride glasses is of special interest. It is known that crystalline fluorides of many metals, such as BaF_2 and PbF_2, have anionic conductivity. Transference numbers in crystalline CaF_2 were determined quite recently by Ure [7], both by the Tubandt method and with the aid of radioactive isotopes. Within the limits of experimental error the transference number of the fluoride ion was found to be unity. We postulated that anionic conductivity persists also in fluoride glasses. Until recently only glasses with cationic or electronic conductivity were known.

Investigation Technique

The experimental glasses were prepared from the appropriate fluoride mixtures by fusion in a platinum crucible in a Silit furnace.* The fusion temperature did not exceed 1000°C. To ensure maximum homogeneity, the melt was agitated by means of a laboratory mixer with a platinum stirrer. Molten fluorides are hydrolyzed fairly easily by atmospheric moisture [8]. Carbon dioxide was passed through the heating zone of the furnace throughout the melting in order to prevent hydrolysis. The molten melt was poured into metal molds heated to 200°C and the glasses were put into a muffle furnace for annealing. The specimens used for the conductivity measurements were in the form of disks 20 mm in diameter and about 5 mm thick. The conductivity measurements were performed by the usual method, described in detail by O. V. Mazurin and A. S. Levin [9].

Results of the Determinations

Two series of glasses were synthesized. The effects of additions of various alkali-metal fluorides to alkali-free glasses were studied in the first series, and the effect of alkaline-earth fluorides, replacing beryllium fluoride in multialkali glass, on the conductivity in the second. Unfortunately, only fluoride glasses of relatively complex composition exhibit a sufficient tendency to glass formation for production of specimens of good quality. It was therefore difficult to expect that the relationships found would be as clear as for the simplest silicate and borate glasses. Nevertheless, we succeeded in demonstrating quite definite relationships. Table 1 gives the composition of the experimental glasses (in moles), and the results of conductivity determinations (logarithms of the resistivity at 150 and 250°C are given). Table 1 also contains values of the activation energy E and of the statistical factor A, found from the known formula:

$$\varkappa = Ae - \frac{E}{2kt}$$

The weak influence of added alkali fluorides on conductivity may first be noted. It is well known that introduction of 15-20% of alkali oxides into any alkali-free oxide glass lowers the resistivity of the glass by 2-6

* M. V. Proskuryakov took part in preparation of the glasses, and N. G. Suikovskaya assisted in the determinations.

TABLE 1. Electrical Parameters and Compositions of Experimental Fluoride Glasses

Glass composition, moles	log ρ 150°C	log ρ 250°C	E, eV	log A
Series 1				
$30BeF_2$, $20AlF_2$, $30CaF_2$, $20SrF_2$, (Gl. No. 30)	12.40	9.35	2.67	3.56
Glass 30 + 5LiF	12.60	9.35	2.85	4.40
Glass 30 + 5NaF	12.42	9.42	2.63	3.26
Glass 30 + 10NaF	12.65	9.47	2.78	3.95
Glass 30 + 15NaF	12.47	9.39	2.70	3.61
Glass 30 + 15NaF (second meeting)	12.16	9.91	2.85	4.82
Glass 30 + NaF	11.55	8.48	2.69	4.48
Glass 30 + 20NaF (second meeting)	11.84	8.69	2.75	4.61
Glass 30 + 5KF	12.08	9.11	2.60	3.44
Glass 30 + 15KF	11.83	8.76	2.69	4.20
Glass 30 + 5CsF	11.65	8.77	2.52	3.40
Glass 30 + 10CsF	11.41	8.71	2.36	2.70
Glass 30 + 15CsF	11.11	8.39	2.38	3.12
Glass 30 + 7.5NaF + 7.5KF	12.40	9.35	2.67	3.56
Glass 30 + 7.5NaF + 7.5CsF	12.16	9.05	2.72	4.08
Glass 30 + 7.5KF + 7.5CsF	11.49	8.73	2.42	2.94
Series 2				
$60BeF_2$, $10AlF_2$, 20CsF	12.92	8.35	4.00	11.01
$60BeF_2$, $10AlF_2$, 20CsF, $10MgF_2$	13.85	11.15	2.36	0.26
$60BeF_2$, $10AlF_2$, 20CsF, $10CaF_2$	15.89	10.81	4.45	10.69
$60BeF_2$, $10AlF_2$, 20CsF, $10SrF_2$	14.36	9.74	4.05	9.81
$60BeF_2$, $10AlF_2$, 20CsF, $10BaF_2$	13.83	9.20	4.05	10.40

orders of magnitude. In this case the greatest decrease of resistivity is less than one and one-half orders of magnitude. The fact that the greatest decrease of resistivity is produced by additions of cesium fluoride is very significant. The conductivity of oxide glasses (both silicate and borate) increases with decreasing radius of the alkali-metal ion in the glass. In the case of fluoride glasses the reverse relationship is distinctly seen (Fig. 1). It is important to emphasize that this relationship also has a fairly definite influence on the activation energy.

Another important and interesting distinction between fluoride and oxide glasses is the total absence of what is known as the two-alkali effect, i. e., the sharp increase of resistivity when one alkali-metal ion is partially replaced by another [10]. It follows from Fig. 2 that the resistivity alters uniformly in the transition from sodium to potassium or cesium glass, or from potassium to cesium glass.

The results of conductivity measurements for the second series of glasses indicate that the resistivity is increased by introduction of 10% of alkaline-earth fluorides. Calcium fluoride increases the resistivity considerably more than barium fluoride, while strontium fluoride occupies an intermediate position (the behavior of magnesium fluoride does not fit into the general relationship). This again is directly opposite to the relationships found for oxide glasses, as the resistance of glasses containing alkali and alkaline-earth oxides increases with increasing radius of the alkaline-earth ion.

Discussion of Results

Special experiments showed that the conductivity of fluoride glasses alters very little on crystallization, mostly within one or two orders of magnitude, and the resistivity of the crystallized specimen is often even lower than that of the glass. This in itself suggests that the charge carriers in the glassy and crystalline speci-

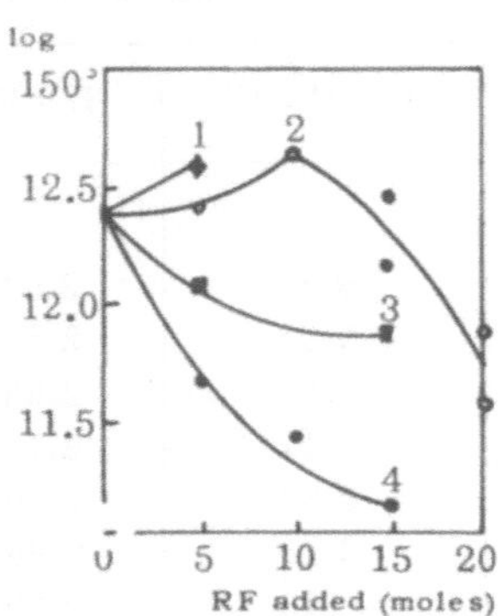

Fig. 1. Effect of added alkali fluorides on the conductivity of glass at 150°C. Additions: 1) LiF; 2) NaF; 3) KF; 4) CsF.

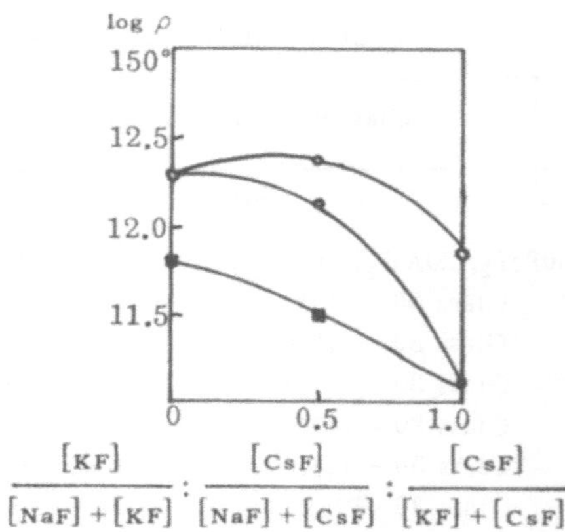

Fig. 2. Effect of replacement of one alkali ion by another on the conductivity of glass at 150°C. The glass contains 15 moles of RF. Replacement: 1) NaF by KF; 2) NaF by CsF; 3) KF by CsF.

men is often even lower than that of the glass. This in itself suggests that the charge carriers in the glassy and crystalline specimens are of the same nature. However, even weightier evidence in favor of the anionic nature of conduction in fluoride glasses is provided by all the experimental data on the influence of composition on the conductivity of these glasses, which can be interpreted only with the assumption that fluoride ions are the charge carriers in fluoride glasses. The weak influence of alkali ions becomes understandable: these ions are not involved in transport of electricity and the change in the conductivity of the glass resulting from introduction of alkali fluorides is due only to the fact that weakly bound fluoride anions are introduced into the glass with these fluorides. The bond between the alkali cation and the fluoride ion weakens with increasing radius of the cation. Therefore, cesium fluoride produces the greatest decrease of resistivity, whereas the first additions of lithium fluoride even increase resistivity (more than 5% of lithium fluoride cannot be introduced into the glass because of very extensive crystallization).

The reason for the absence of the two-alkali effect also becomes apparent. If the current in the glass is carried by anions, a sharp increase of resistivity can occur only if some of the anions are replaced by anions of a different kind. It is evident that if we want to observe an effect analogous to the two-alkali effect in a glass with anionic conductivity, we must introduce a substance in which the charge carriers are anions of a different type (such as $PbCl_2$). We have as yet been unable to find a substance which is an anionic conductor, not a fluoride, and readily compatible with fluoride glass.

Let us consider in greater detail the influence of bivalent-metal fluorides on conductivity. The decrease of the conductivity of alkali silicate glasses on introduction of bivalent-metal oxides is explained in a paper by O. V. Mazurin and G. T. Petrovskii [11].

They emphasize that alkali-metal ions move through silicate glasses in a medium containing oxygen anions, and the mobility of the cations is determined primarily by the strength of their bonds with the oxygen environment. The bonding between the alkali and the oxyge ions becomes stronger with weaker polarization of the oxygen ions by other ions present in the glass. The polarizing effect of a cation increases with increase of its charge and decrease of its radius. Therefore, introduction of bivalent-metal oxides into glass should result in increase of resistivity, which becomes greater with increase of the size of the cation introduced.

We must stress once again that bivalent cations introduced into alkali silicate glass influence the mobility of the charge carriers through an intermediary, the oxygen ion, and not directly. Fluoride glasses contain no such intermediaries, and the positive cations have a direct influence on the charge carriers. Naturally, the smaller the size and the greater the charge of a cation, the more strongly does it attract fluoride ions, and the

TABLE 2. Resistivity Parameters of Certain Typical Oxide Glasses

Glass compositions, mole %	log ρ 150°C	log ρ 250°C	E, eV	log A
$80SiO_2$, $20Na_2O$	6.40	4.80	1.40	1.96
$60SiO_2$, $20Na_2O$, $20K_2O$	8.11	5.76	2.05	4.18
$60SiO_2$, $20Na_2O$, 20BaO	9.30	7.31	1.93	1.14
$90B_2O_3$, $10Na_2O$	14.46	11.36	2.72	1.8
$80B_2O_3$, $20Na_2O$	9.75	7.57	1.92	1.9
$60SiO_3$, 40PbO	12.19	9.61	2.26	1.3
$60B_2O_3$, 40BaO	18.17*	14.42*	3.28	1.4
$60B_2O_3$, $10Al_2O_3$, 30BaO	20.52*	16.0*	3.97	3.1

*Extrapolated from higher temperatures.

more the movement of the charge carriers is hindered. From this standpoint it is quite easy to understand the comparative effects of different alkali fluorides and also the changes of resistivity on introduction of calcium, strontium, and barium fluorides.

The peculiar behavior of magnesium fluoride and beryllium fluoride is noteworthy. When fluoride ions strongly bonded to beryllium are replaced by fluoride ions weakly bonded to calcium, strontium, or barium, the resistivity of the glass increases. When BeF_2 is replaced by MgF_2 the resistivity increases to approximately the same extent as when BeF_2 is replaced by BaF_2. To explain these discrepancies, we must again turn to [11], where it is shown that the ionic radius rule is valid only for cations having the same coordination number. Calcium, strontium, and barium fluorides have a fluorite structure; magnesium fluoride has a rutile structure; beryllium fluoride is isomorphous with cristobalite. Naturally, there cannot be a general relationship in this case, because of the different degrees of shielding of the cation by the anion in different structures. In the fluoberyllate tetrahedron the fluoride ions shield the cation completely and its field does not act on the fluoride ions moving in the glass. Large ions, because of their high coordination number, can interact with fluoride ions introduced into the glass with other fluorides. As an example, let us consider the changes taking place in glass on partial replacement of beryllium fluoride by calcium fluoride. This results in some increase in the amount of mobile fluoride ions in the glass, but the resistivity of the glass increases instead of decreasing, because a cation which is quite inert toward the mobile ions has been replaced by a cation capable of interacting with the mobile cations and thus lowering their mobility. The action of magnesium ions, as was to be expected, is intermediate between the effects of the other alkaline-earth ions and beryllium. We do not attempt at present to explain the low activation energy of magnesium glass.

In conclusion, it is interesting to compare the parameters characterizing resistivity of fluoride glasses with the corresponding parameters for oxide glasses. Table 2 contains data on certain typical oxide glasses.

Fluoride glasses of the first series with low contents of the network former (beryllium fluoride) resemble lead silicate glasses, but have somewhat higher activation energy and a greater statistical factor.

With regard to glasses of the second series, there are no analogous compositions at all among the oxide glasses. The activation energies of these fluoride glasses approach and in some cases exceed the activation energy of barium aluminoborate glasses, which have the highest activation energy of all known oxide glasses. At the same time, the resistivity of fluoride glasses at 100-300°C is lower by 4-6 orders of magnitude than that of borate glasses. This is due to the enormous value of the statistical factor. The highest value of log A found for oxide glasses is 4.7. For the great majority of oxide glasses log A ranges from +2.5 to −1, in good agreement with theory. The values of log A for fluoride glasses of the second series are in the range of 10-11. Elucidation of the causes of this effect is of undoubted interest, but much experimental data most be collected before this problem is studied.

From the practical standpoint, it may be possible to use fluoride glasses as insulating materials of high resistivity and exceptionally high temperature coefficients of conductivity, with low softening temperatures

(400-500°C) and very large expansion coefficients. The coefficients of expansion of some of the glasses described in this article are 155-160 · 10^{-7}, which is very close to the coefficient of expansion of copper.

Literature Cited

1. G. Heyne, Angew. Chem. 46:(28):473 (1933).
2. Minora Imaoka and Shinga Mizusawa, J. Ceram. Assoc. Japan 62(64):38 (1954).
3. M. S. Genrikh and L. I. Ignat'eva, Optiko-mekhanicheskaya promyshlennost', No. 6:46 (1957).
4. K. Gerth and W. Vogel, Silikat Tech. 9(8):353 (1958); 9(11):495 (1958).
5. L. R. Batsanova, A. V. Novoselova, et al., Izv. Vuzov (Khim. i Khimichesk.Tekhnol.,No. 5:751 (1959).
6. T. Izumitani and R. Terai, Bull. Osaka Ind. Res. Inst. 3(1):21, 104 (1952).
7. R. Ure, J. Chem. Phys. 26(6):1363 (1957).
8. N. A. Toropov, M. M. Sychev, and E. D. Alekseeva, Tr. Leningr. Tekhnol. Inst. im. Lensoveta, No. 34 (1955).
9. O. V. Mazurin and A. S. Levin, Izv. Vuzov (Khim. i Khimichesk. Tekhnol.), No. 2:142 (1958).
10. O. V. Mazurin and E. E. Borisovskii, Zh. Tekhn. Fiz. 27(2):275 (1957).
11. O. V. Mazurin and G. T. Petrovskii, Tr. Leningr. Tekhnol. Inst. Lensoveta, p. 51 (1956).

Received July 26, 1960.

NATURE OF THE CONDUCTIVITY OF SODIUM ALUMINOSILICATE GLASSES

R. L. Myuller and A. A. Pronkin

Introduction of aluminum oxide into alkali silicate glasses is known to produce considerable increase of dielectric constant and losses, and of conductivity.

The higher losses in alkali aluminosilicates can be satisfactorily explained by relaxational displacements of bonded cations in $Al^{-}O_{4/2}$ units [1, 2]:

```
--O   O--              --O   O--
   \⊖/                     \⊖/
 M⊕ Al        ⇌            Al  M⊕
   /  \                    /  \
--O    O--             --O    O--
```
(1)

The content of polar $M^{+}Al^{-}O_{4/2}$ units is increased on introduction of Al_2O_3 into an alkali glass, and this should increase the dielectric constant, conductivity, and losses. This increase can continue until the Al_2O_3 added has absorbed all the alkali-metal oxide in the glass ([Al]/[M]= 1). Further addition of Al_2O_3 is accompanied by formation of nonpolar $AlO_{3/2}$ units with the closest packing of oxygen atoms. This increased packing density results in an increase of the energy which must be expended by the cations to overcome the repulsive forces of the electron shells of the atoms during dissociation of polar $M^{+}Al^{-}O_{4/2}$ units. This accounts for the increase of the energy ε_{σ} observed at [Al]/[M] > 1 [3] and the consequent decrease of conductivity in alkali aluminosilicate glasses. The conductivity may also decrease as the result of a decrease of the polar group content when one $SiO_{4/2}$ group is replaced by two $AlO_{3/2}$ groups.

The electron resonance concept has also been put forward to explain the "anomalous" electrical properties of alkali aluminosilicate glasses; dielectric losses in these glasses at 200-500 K are attributed to resonance displacements of electrons in the tetrahedral $Al^{-}O_{4/2}$ units:

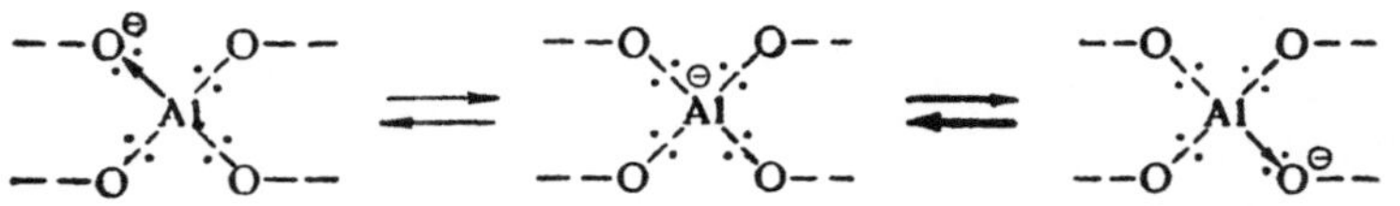

```
--O⊖  O--        --O   O--        --O   O--
    \:            \⊖/              \  :
     Al     ⇌      Al      ⇌        Al
    : \           / \              :  \
--O    O--     --O   O--        --O    O⊖--
```
(2)

mixed electronic–ionic conductivity is postulated in this temperature range [4].

The object of the present work was direct experimental verification of the nature of the conductivity in alkali aluminosilicate glasses.

Transition from ionic to electronic conductivity in the specified temperature range should be accompanied by considerable changes in the values of σ_0 and ε_{σ} in the expression

$$\sigma = \sigma_0 \exp\left(-\frac{\varepsilon_\sigma}{2kT}\right) \tag{3}$$

The structural unit $M^{+}Al^{-}O_{4/2}$ is the source of current carriers (M^{+} ions and electrons). For a concentration $[M^{+}Al^{-}O_{4/2}]=n$ moles/cm^3, the theoretical value of the conductivity modulus, both for ionic and for electronic conductivity, is $\log(\sigma_0 T/n) = 4 \pm 1$ [5].

Glass No. (number of specimens)	Glass composition						Mol. wt., M	Density d, g/cm³	Conc. [M], moles/cm³ 10^3	ε_σ, eV	$\log \sigma_0$	$\log \sigma_0/x[M]$
	Molar percentages			Average chemical structural unit $M = (Na^+Al^-O_{4/2})_x(Na^+O^-SiO_{3/2})_{1-x}(SiO_{4/2})_y$								
	Na_2O	Al_2O_3	SiO_2	x	$1-x$	y						
1 (1)	30.0	10.0	60.0	0.33	0.67	0.33	108.0	2.47_4	2.28	1.33	2.35	4.5
2 (2)	20.0	10.0	70.0	0.50	0.50	1.25	161.6	2.41	1.50	1.41	2.00	4.1
3 (1)	25.0	10.0	65.0	0.40	0.60	0.70	129.3	2.44_6	1.90	1.25	1.13	3.3
4 (4)	30.0	20.0	50.0	0.67	0.33	0.49	114.6	2.46_3	2.15	1.35	2.5	4.3
5 (2)	25.0	20.0	55.0	0.80	0.20	0.90	137.7	2.45_4	1.78	1.30	2.37	4.2
6 (2)	25.0	15.0	60.0	0.60	0.40	0.80	133.6	2.45_2	1.84	1.37	2.71	4.7
7 (4)	30.0	15.0	55.0	0.50	0.50	0.42	111.5	2.48	2.22	1.37	2.4	4.4
										Mean		4.2 ± 0.3

In the temperature region of transition from electronic to ionic conductivity, entropy of conductivity $\Delta S_\sigma = -\Delta\varepsilon_\sigma/\Delta T$ should exist, and higher values of $\sigma_{0e} = \sigma_{0T} \exp(\Delta S\sigma/2k)$ [6] and therefore of the conductivity modulus $\log(\sigma_{0e}/n) = \log[\sigma_{0T} \exp(\Delta S_\sigma/2k)/n] \gg 4 \pm 1$ should be observed.

If this modulus remains unchanged [$\log(\sigma_{0e}/n) = 4 \pm 1$] in the temperature range in question, we have ε_σ constant, and the possibility of transition from the one type of conductivity to the other is thereby excluded. Ionic conductivity at high temperatures would then indicate that the conductivity of alkali aluminosilicate glasses is of the ionic type throughout the temperature range in question.

The compositions and properties of the glasses investigated are given in the table. Electrodes of Na amalgam were used for the conductivity measurements (with some glasses copper cathodes were used). The table shows that in the "transitional" region of 15-110°C a constant value was found for the conductivity modulus, $\log(\sigma_0/n) = 4.2 \pm 0.3$, which is in agreement with the theoretical value. All this shows that the nature of the current carriers remains unchanged in this temperature region.

Transference numbers were then determined by Tubandt's method. About 20 Cb was passed through the glass, the current was cut off, and the furnace was cooled during 10-12 hr to room temperature.

Calculations of the transference number η_{Na} from the experimental data obtained by Tubandt's method for the sodium aluminosilicate glasses at 260-300°C gave 0.996 ± 0.005 for the transference number of Na ions.

This result, together with the agreement between the experimental and theoretical values for the conductivity modulus, confirms conclusively the ionic conductivity of these aluminosilicate glasses in the region of 300-570°K.

Literature Cited

1. R. L. Myuller, Zh. Tekhn. Fiz., 25:1556, 1967 (1955).
2. V. I. Odelevskii and N. M. Verebeichik, Izv. Tomsk.Politekhn. Inst., 91:247 (1956).
3. J. O. Isard, J. Soc. Glass Technol., 43:113 (1959).
4. V. A. Ioffe and I. S. Yanchevskaya, Zh. Tekhn. Fiz., 28:2154 (1958); Collection: Physics of Dielectrics, Izd. Akad. Nauk SSSR (1960) pp. 182, 218; Fiz. Tverd. Tela, 4:676 (1962).
5. R. L. Myuller, Zh. Prikl. Khim., 35:541 (1962).
6. R. L. Myuller, Zh. Fiz. Khim., 6:616 (1935); Fiz. Tverd. Tela, 2:1345 (1960); Acta Physicochim. USSR, 2:103 (1935); Zh. Tekhn. Fiz., 25:236, 246, 2428 (1955).

INFLUENCE OF GAMMA RADIATION ON THE STRUCTURE OF GLASS

L. M. Belyavskaya

Investigations of the effects of gamma radiation on the conductivity of sodium silicate glasses in strong electric fields revealed a number of effects not observed in glasses in the normal state [1, 2]. To elucidate the nature of these changes due to radiation, we studied the effect of temperature at field strengths $E < E_{cr}$, where E_{cr} is the limit of applicability of Ohm's law.

The galvanometer circuit used by us earlier [3] was used for the determinations. The results discussed below were obtained after gamma irradiation, dose $1.6 \cdot 10^6$ r, $h\nu = 1.25$ MeV, $E = 83{,}000$ V/cm. Specimens which showed no appreciable changes of conductivity during the period of observation at room temperature were chosen for the experiments. However, during and after heating the return to the original state was greatly accelerated for all the specimens; it was therefore necessary to use the maximum and minimum values of the quantities measured. The determinations were therefore carried out with both rising and falling temperatures (see arrows in the figure).* On further heating (to 250°C) the current increases so much that the cooling curve can be obtained only at lower temperatures.

The curve obtained with falling temperature gives a high value for the free energy of electrolytic dissociation [4] ($\Psi = 11$-37 kcal/mole) while the rising-temperature curve gives a low value ($\Psi = 2.7$ kcal/mole†). The normal value of Ψ before irradiation (at the given sodium ion concentration [M] = 2.3 moles/ml) is 30 kcal/mole [4, 5, 6, 10].

At temperatures close to 100 and 172°C the current passing through the specimen becomes unstable; this effect is not observed before irradiation. It is probable that at these points silica undergoes polymorphic transitions, detected earlier by various workers [7, 8, 9] in glasses of this composition. Irradiation may help to reveal effects of this type, associated with first- or second-order phase changes [7, 11].

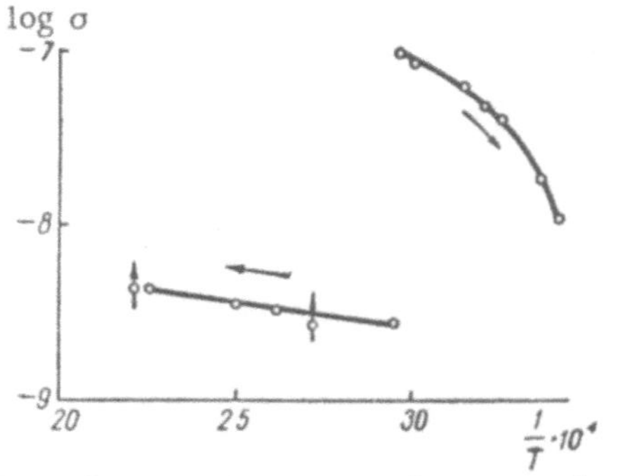

Effect of temperature on conductivity of a specimen subjected to gamma radiation ($1.6 \cdot 10^6$ r), with rising and falling temperatures at field strength $E = 83{,}000$ V/cm ($E < E_{cr}$).

The decrease of free energy of electrolytic dissociation and its gradual recovery after heating at about 200°C is apparently characteristic for glasses of this composition after sufficient doses of gamma irradiation. The process associated with increase of Ψ and the corresponding decrease of current occurs at current values higher by 1-2 orders of magnitude. In some cases it is associated with restoration of high-voltage polarization after irradiation [10]. However, this restoration cannot cause the observed decrease of Ψ. In this case, as in analysis of concentrational relationships [10], it is convenient to postulate association of polar structural groups, which probably play the leading part in electrical conduction at high Na_2O contents (in the present instance 26 mole %) [4].

*The left-hand branches of the curves cannot be obtained, as partial breakdown (an irreversible state of the glass) occurs under such conditions.

†The values of Ψ were determined from the angle formed by the curve with the axis of abscissas.

Therefore it appears that gamma irradiation increases the average size of the structural regions; this is consistent with our quantitative estimates carried out by an independent method in studies of the conductivity of irradiated glasses in strong fields [1, 2].

It is very important, in our opinion, to verify this hypothesis by x-ray diffraction investigations of specimens before and after irradiation.

Literature Cited

1. L. M. Belyavskaya, Summary Proceedings of Conference on the Properties of Materials under the Influence of Radiation, Tomsk (1961).
2. L. M. Belyavskaya, Izv. Vuzov, Fizika (1) (1963).
3. L. M. Belyavskaya, Izv. Vuzov, Fizika (1) (1959).
4. R. L. Myuller, Zh. Tekhn. Fiz., 25:346 (1955).
5. R. L. Myuller, Zh. Tekhn. Fiz., 25:336 (1955).
6. R. L. Myuller, Zh. Tekhn. Fiz., 25:1868 (1955).
7. V. A. Presnov, V. I. Gaman, and L. M. Krasil'nikova, Collection: The Glassy State, Izd. Akad. Nauk SSSR (1960). [English translation: The Structure of Glass, Vol. 2, Consultants Bureau, New York (1960).]
8. A. M. Venderovich and L. M. Krasil'nikova, Izv. Vuzov, Fizika (4) (1958).
9. L. M. Belyavskaya, Collection: The Glassy State, Izd. Akad. Nauk SSSR (1960). [English translation: The Structure of Glass, Vol. 2, Consultants Bureau, New York (1960).]
10. R. L. Myuller, Zh. Tekhn. Fiz., 26:2614 (1955).
11. V. I. Gaman, Author's summary of candidate's dissertation, Tomsk (1958).

THERMOELECTRIC EFFECTS AND STRUÇTURE OF SODIUM SILICATE GLASSES

A. F. Borisov and V. I. Zadumin

A new trend in the physical chemistry of silicate melts is the use of cells of the type

$$\text{Pt} \mid \text{melt I} \mid \text{melt II} \mid \text{Pt} \quad (1)$$

for studying the molecular state of melts.

In our country, the first work in this field was carried out under Dertev's guidance at the A. A. Zhdanov Gor'kii Polytechnic Institute [1].

In this paper we examine the use of cells of type (1) for studying sodium silicate melts.

Experimental

The following technique was devised for studying thermoelectric effects in melts (Fig. 1). A platinum boat 2 containing crushed glass was inserted into the furnace 1, heated to the experimental temperature. After the glass had melted, platinum electrodes 3 were immersed in it, with platinum–platinum-rhodium thermocouples in their immediate vicinity. The cooling device 4, consisting of a copper U-shaped tube through which water was circulated, was inserted into the furnace to create a temperature gradient.

The mechanism whereby a thermoelectromotive force arises between platinum electrodes in sodium silicate melts may be represented as follows. Oxygen ions in the glass undergo thermal vibrations about definite centers of equilibrium, and when these vibrations are intense oxygen ions can migrate from one silicon atom to a neighboring one.

Accordingly, the electrochemical activity of oxygen is determined by a quantity proportional to another in a definite volume of the glass melt.

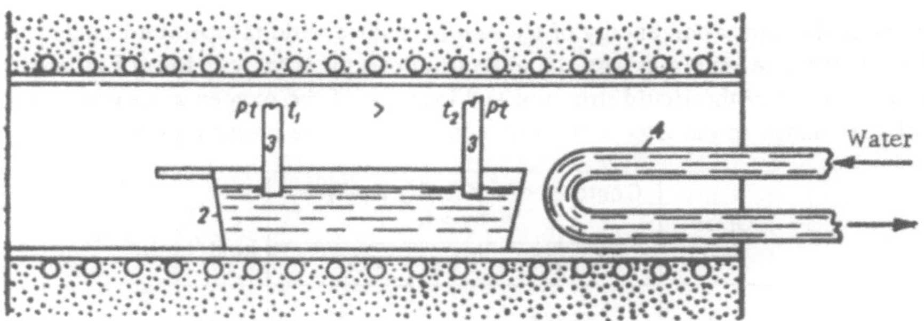

Fig. 1. Line diagram of apparatus for studying thermoelectric effects in melts. 1) Furnace; 2) platinum boat; 3) platinum electrode; 4) copper tube.

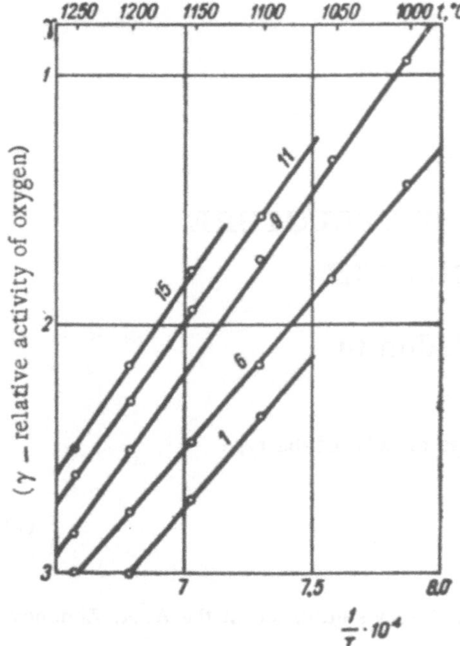

Fig. 2. Relative activity of oxygen as a function of the temperature (the numerals on the plots correspond to the glass numbers).

In the boat with a temperature gradient the oxygen activity differs in different regions. The relative activity may be measured by means of platinum electrodes, the reversibility of which with respect to oxygen ions was confirmed by the work of Minenko et al. [2].

The thermo-emf is given by the following expression:

$$E = \frac{RT_1}{nF} \ln a' - \frac{RT_2}{nF} \ln a'' \quad (2)$$

where E is the thermo-emf, R is the gas constant, n is the number of electrons involved in the reaction (four in this case), and a' and a" are the oxygen-ion activities at temperatures T_1 and T_2.

Equation (2) can be used for calculating the relative activity of oxygen at various temperatures, the activity of oxygen in the glass melt at a certain temperature being taken as unity.

The oxygen-ion activation energy is calculated from the expression

$$\gamma = Ke^{-\frac{U}{RT}} \quad (3)$$

where γ is the relative activity of oxygen ions at temperature T, K is a constant, and U is the average activation energy of the oxygen ions.

Our measurements show that plots of log γ versus 1/T for various glasses of the system Na_2O-SiO_2 are straight lines (Fig. 2), the slopes of which are a measure of the activation energy of the oxygen ions (the glass compositions and activation energies are given in the table— see below).

Evaluation of the data obtained by Johnson et al. [5] on the activation energy of sodium ions, and of our results on the activation energy of oxygen ions, gives a complete picture of the structure and energy parameters of the system Na_2O-SiO_2.

Comparison of the structural and energy characteristics, the energy of activation of viscous flow, and the temperature shows that the activation energy of viscous flow depends on the activation energies of oxygen and sodium ions and on the quantitative ratio of Si—O and O—Na bonds, which is determined by the composition of the glass; this confirms the theoretical views of Evstrop'ev [3] and Slavyanskii [4].

The constancy of the activation energy of Na ions [5] over a wide temperature range, and of oxygen ions above the liquidus temperature, explains why the activation energy of viscous flow of glasses is constant at high temperatures. Complication of the liquid structure and increase of the oxygen activation energy are probably the main causes of the change in the flow activation energy below the melting point.

Glass No.	Contents of oxides by analysis, % by wt.		U, kcal/mole
	Na_2O	SiO_2	
1	58.68	41.32	57
6	35.98	64.02	57
9	32.97	67.03	71
11	28.80	71.20	73
15	19.97	80.03	73

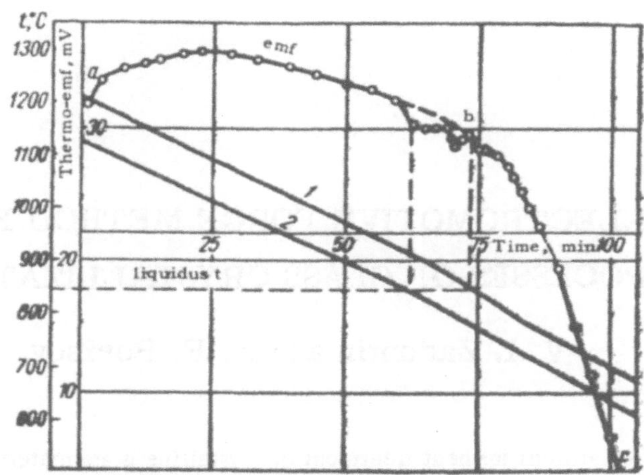

Fig. 3. Effect of temperature on emf for glass No. 6. 1) Temperature of first electrode; 2) temperature of second electrode.

The change in the activation energy of oxygen ions can be deduced from Fig. 3, which presents the effect of temperature on the thermo-emf of glass 6. The temperature gradient between the electrodes was constant.

The curve can be divided into the regions ab and bc.

The dependence of thermo-emf on temperature in region ab can be calculated from Eq. (2); the region bc represents changes of thermo-emf during structural changes in the melt.

The high sensitivity of platinum–oxygen electrodes to structural changes in melts offers a good basis for development of a dynamic method for studying crystallization and vitrification processes.

Literature Cited

1. A. F. Borisov and N. K. Dertev, Tr. Gor'kovsk. Politekhn. Inst. im. A. A. Zhdanova, Khimiko-Tekhnologicheskii i Silikatnyi Fakul'tet, 13 (5):16 (1957).
2. V. I. Minenko, S. M. Petrov, and S. I. Ivanova, Izv. Vuzov, Chernaya Metallurgiya (7):10 (1960).
3. K. S. Evstrop'ev, Collection: Physicochemical Properties of the Ternary System $Na_2O-PbO-SiO_2$, Izd. Akad. Nauk SSSR (1949).
4. V. T Slavyanskii, Collection: The Glassy State, Izd. Akad. Nauk SSSR (1960). [English translation: The Structure of Glass, Vol. 2, Consultants Bureau, New York (1960).]
5. J. R. Johnson, R. H. Bristow, and H. H. Blau, J. Am. Ceram. Soc., 34 (6) (1951).

USE OF AN ELECTROMOTIVE FORCE METHOD FOR STUDYING PROCESSES OF GLASS CRYSTALLIZATION

V. I. Zadumin and A. F. Borisov

The development of conditions for heat treatment of crystalline glass materials is hindered by the lack of a method for observing changes in the relative proportions of crystalline and glassy phases directly during the heat treatment.

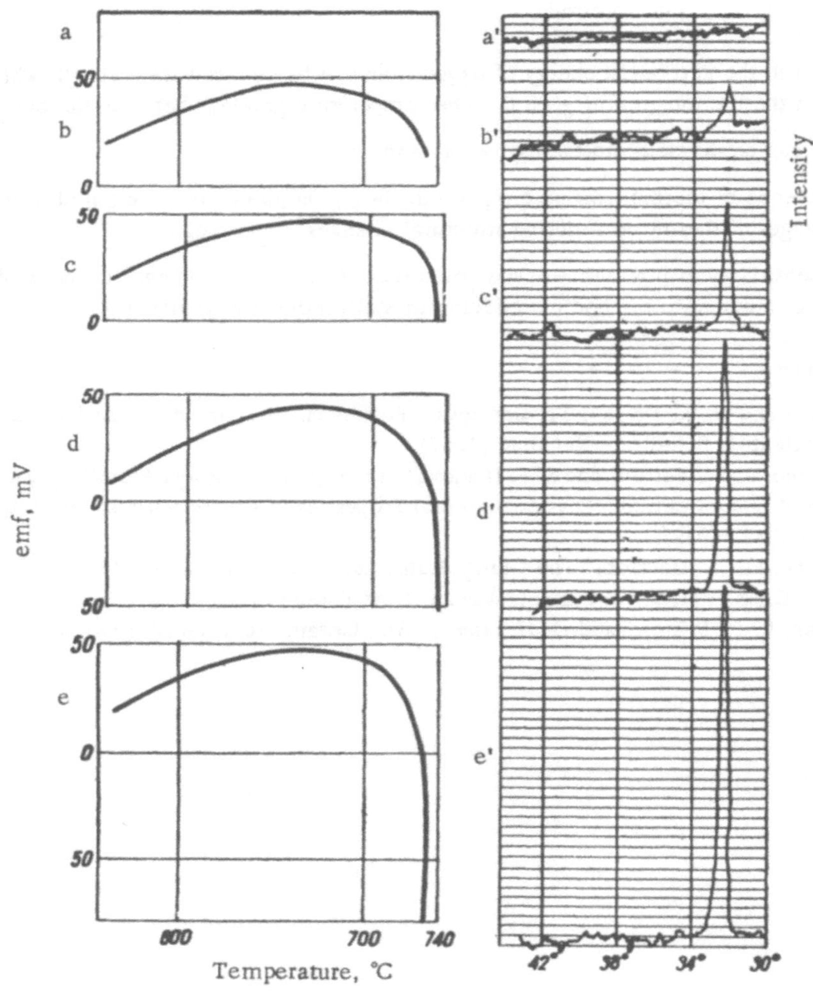

Fig. 1. Effect of heat treatment on the emf and the height of the maximum at 32°30'. a) Glass before heat treatment; b, c, d, e) heat-treated glasses; a') x-ray diffraction trace of glass before heat treatment; b', c', d', e') x-ray diffraction traces of heat-treated glasses.

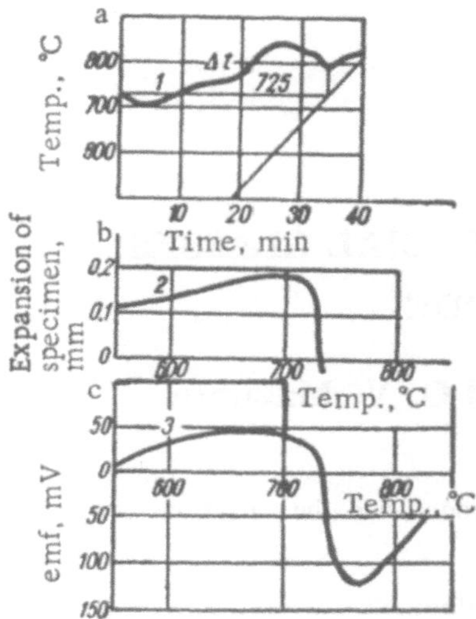

Fig. 2. Results of investigations of glass crystallization by different methods. a) Differential thermal analysis; b) expansion of the specimen during heat treatment; c) variations of emf with temperature.

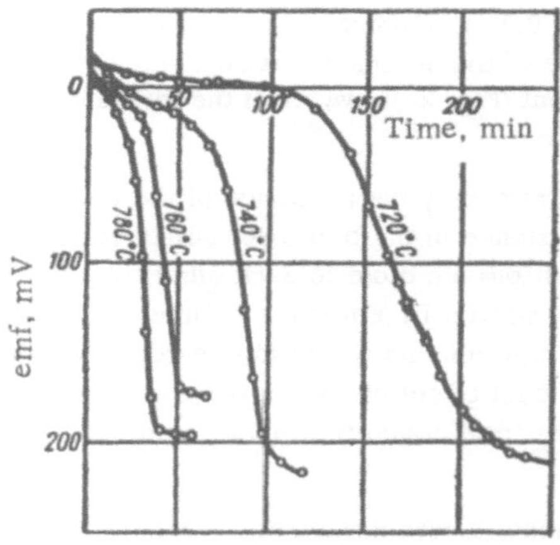

Fig. 3. Kinetics of glass crystallization at various temperatures.

The Saratov Branch of the Institute of Glass has devised a method for continuous investigation of crystallization kinetics; the method is highly sensitive to structural changes in glass.

The method is based on the following principle: two specimens are placed between two platinum electrode plates. One specimen is a reference standard which does not undergo structural changes. A cerium dioxide plate 1-1.5 mm thick is used for this purpose. The plate is molded from cerium dioxide powder and fired at 1400-1500°C. The second specimen is a plate of the glass under investigation, 0.5-1.0 mm thick. The system is placed in a furnace and heated with the furnace. The emf between the platinum electrodes is measured during the heating.

The results obtained in a study of crystallization processes by the emf method are compared below with the results of x-ray structural, differential thermal, and dilatometric analyses. Glasses of the system $Li_2O-Al_2O_3-SiO_2$ with added titanium dioxide were investigated.

The system was heated at the rate of 2-3° per minute to 730°C, which is the optimum crystallization temperature for this glass. The specimens were held for a definite time at 725-730°C. The emf was measured at the same time. When a certain emf was reached the specimens were suddenly chilled; a definite ratio of crystalline and glassy phases resulting from the heat treatment was thus frozen in the specimen.

The experimental results in Fig. 1 show that the emf of the system undergoes sharp changes during crystallization; a steady increase of the amount of crystalline phase corresponds to a fall of emf.

The height of the maximum accordingly increases progressively from specimen b' to specimen e'

The results of differential thermal and dilatometric analyses are compared in Fig. 2 with the emf variations with temperature for the same glass. Curve 1 shows a weak exotherm at 725°C, due to crystallization of the glass.

The specimen contracts sharply at 730°C as the result of crystallization processes in it.

For comparison, curve 3 in Fig. 2 represents variations in the emf of the system during continuous heating of the specimen (heating rate 2° per minute). Figure 2 shows good agreement between the results of differential thermal analysis and the emf method. The emf method records the increasing content of the crystalline phase and permits continuous study of the kinetics of the process.

Figure 3 presents the crystallization kinetics of the glass at various temperatures. The crystallization rate increases continuously with temperature.

POLARIZATION OF HIGH-ALKALI COMMERCIAL GLASSES ADJACENT TO AN ELECTRODE

V. B. Brailovskii, M. Ya. Rozenblyum, and O. V. Mazurin

A direct voltage was applied to a glass specimen with graphite electrodes, and the variations of current with time were investigated.

The decrease of current with time in the case of polarization at the electrode conforms to a general relationship for different glasses (see the table). In the first approximation the resistivity curve, in the form of a plot of log ρ versus log τ, where τ is time, is made up from two straight lines, one almost parallel to the abscissa and the other forming an angle of 40-50° with that axis. Variations of resistivity of the glasses with time at various temperatures are shown in Fig. 1. For all the glasses studied the quantity of electricity passing through the specimen at 220 V and any temperature before the break in the curve is the same, about $1.5 \cdot 10^{-3}$ Cb/cm^2. The position of the sloping portion of the curve alters little with changes in composition. For glass specimens of the same composition and subjected to the same heat treatment the sloping portions of all the curves lie on the same straight line, with deviations not exceeding 0.1 of an order of magnitude (Fig. 2). Variations of the direct voltage applied to the specimen alter the slope of the inclined portion of the curve somewhat. When the applied voltage is increased during the experiment (Fig. 2, curve ABC) the resistance of the glass drops sharply at first and then rises to approach the initial line.

It is possible that the horizontal component of the curve is determined by the resistance of the main bulk of the specimen while the sloping portion represents the increasing resistance in the thin layer adjacent to the anode. As the sloping portions of the curves for different glass compositions are close to each other, it is likely that the low-conducting layers formed on the different glasses are very similar in properties. Therefore, partial replacement of one alkali-metal oxide by another, or replacement of calcium and magnesium oxides by barium oxide, while changing the resistivity of the glass sharply, has no significant effect on the movement of alkali ions in the region adjacent to the electrode. It should be noted that the total contents of alkali oxides were roughly the same in all the glasses.

The following equation is derived from these results for approximate determination of the effect of time of action of the applied field on the resistivity of glass specimens:

$$\rho_\tau = \rho_{init} + A\tau^x$$

where A is the time constant, close to $3 \cdot 10^5$, which varies somewhat with the composition of the glass; τ is

Composition of the Glasses Studied, % by Weight

Glass code	SiO_2	B_2O_3	Al_2O_3	CaO	MgO	BaO	Na_2O	K_2O
S-89-6	69.5	2.0	—	5.5	3.5	2.0	11.0	6.5
Window	71.0	—	1.5	8.0	3.5	—	15.0	—
No. 30	68.0	2.0	—	—	—	15.0	5.0	10.0

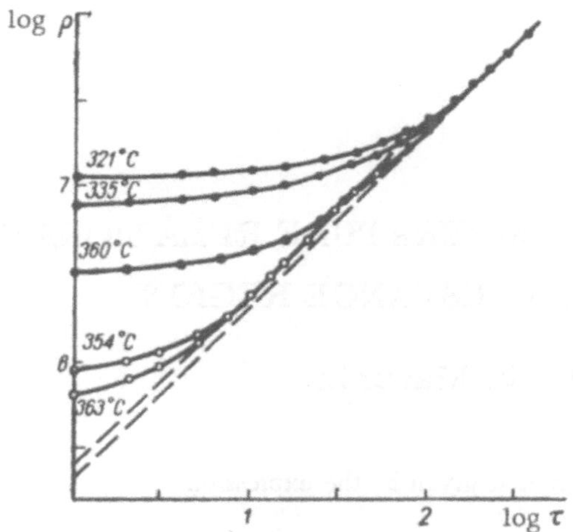

Fig. 1. Variations of the resistivity of various glasses with time at different temperatures. ●) Glass No. 30, 220 V; ○) glass S-89-6, 220 V.

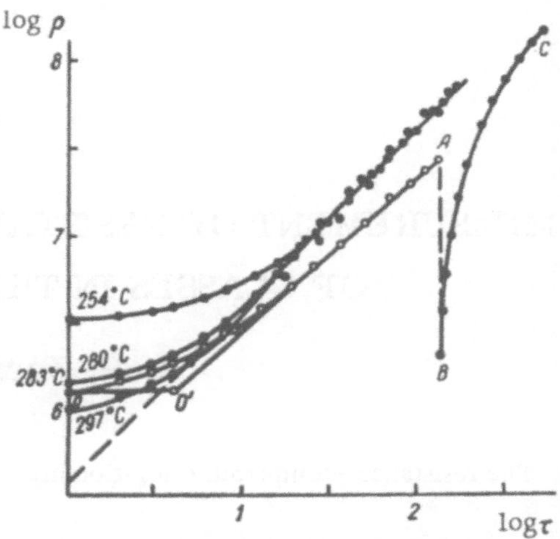

Fig. 2. Variations of the resistivity of window glass with time, at: ●) 220 V; ○) 25 V, the voltage was raised to 220 V at the point A).

the time; x is the slope factor, which depends somewhat on the voltage but is close to unity. This calculation method could be refined by further investigations.

These relationships can probably also be used for determination of the proportions of ionic and electronic conduction in glasses of mixed conduction.

MEASUREMENT OF RESISTANCE–TEMPERATURE RELATIONSHIPS OF GLASSES IN THE HIGH-RESISTANCE REGION

A. P. Zorin and O. V. Mazurin

The resistance–temperature relationship for solid glasses is given by the expression

$$\rho = Ae^{\frac{B}{T}} \tag{1}$$

where A and B are constants, T is the absolute temperature, and e is the base of natural logarithms.

Up to $\rho \simeq 10^{14}$ ohm-cm, Eq. (1) is in good agreement with experimental data. At $\rho > 10^{14}$ ohm-cm many authors have observed deviations from this relation [1]. It should be pointed out, however, that in many of the investigations in this field the techniques used were unreliable. Errors in measurements of high resistances may be due to 1) inadequate insulation of the measurement circuit, 2) influence of absorption currents, or 3) surface conduction.

Bridge methods and capacitor charging methods are generally used for the measurements. One variety of the latter is Townsend's method [2]; with this method the insulation requirements in the measuring part of the circuit are far less stringent and the influence of surface conduction is diminished.

Absorption current can be subdivided into current due to transition processes in the measurement circuit and the polarization current of the dielectric. Transition processes are significant only in measurements by the bridge method. Analysis of the equivalent circuit for this method shows that the time constant $\tau = R_0 \cdot R_X (C_X + C_0)/(R_X + R_0)$ (see Fig. 1) becomes comparable with the time between application of the potential to the specimen and the measurement when the standard resistance $R_0 \geq 10^{10}$ ohms. Therefore measurement of resistances higher than 10^{13} ohms by this method may lead to errors.

Electrical conduction is the passage of a current, unchanging in time, under the action of an unchanging potential through the specimen. Therefore for correct measurement it is necessary to wait until current flow

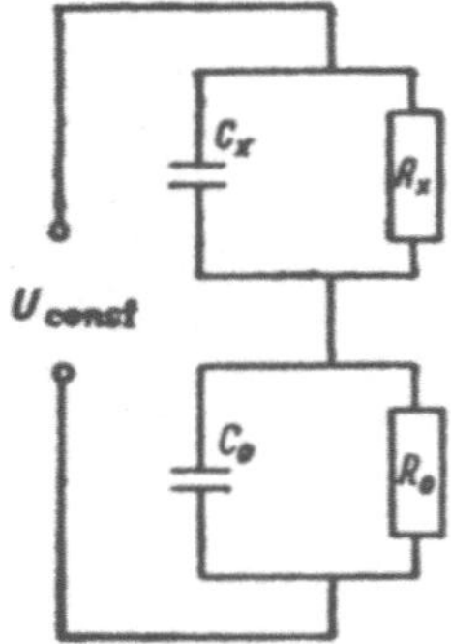

Fig. 1. Equivalent circuit for conductivity measurement by the bridge method.

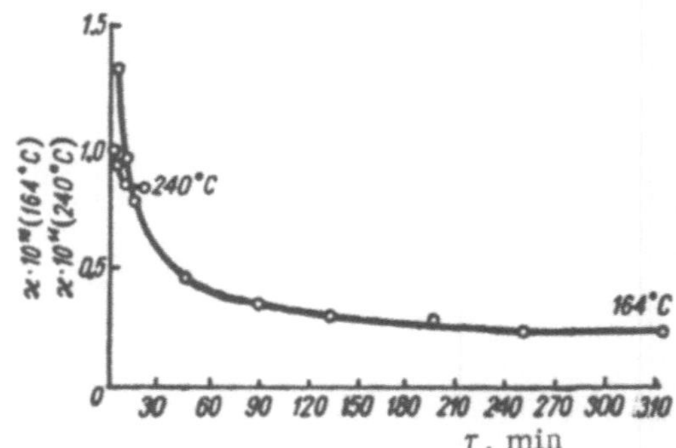

Fig. 2. Conductivity as a function of time for glass No. 1 at two different temperatures.

Compositions of the Glasses Investigated (mole %)

Oxides	Glass				Oxides	Glass			
	№ 15	№ 1	90 K	279-2		№ 15	№ 1	90 K	279-2
SiO_2	60.6	55.6	56.33	42.82	ZrO_2	—	5.55	—	2.35
CaO	30.3	16.65	19.22	9.40	MgO	—	—	6.17	—
Al_2O_3	9.1	—	8.53	0.70	B_2O_3	—	—	8.93	14.11
SrO	—	5.55	—	2.35	PbO	—	—	—	4.70
BaO	—	16.65	—	21.22	F_2	—	—	0.81	—

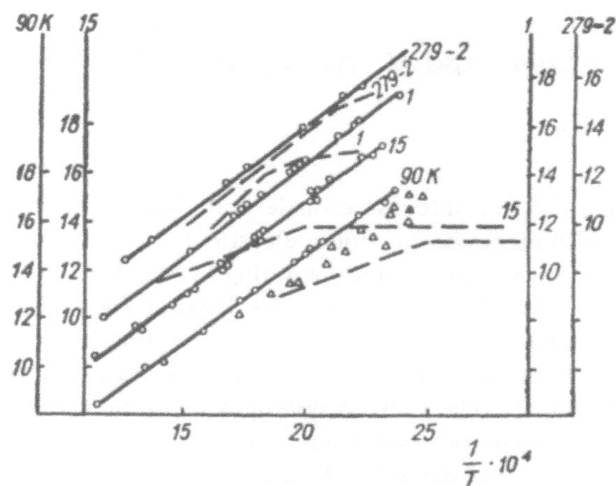

Fig. 3. Variations of resistance with temperature for various glasses. ——) Our data; -------) literature data. The compositions are given in the table. Glass 279-2 No. 1, data from [4]; glass No. 15, data from [5]; standard glass for glass fibers, data from [6].

due to thermal ionic polarization has ceased. The time required for this depends greatly on the temperature (Fig. 2). An essential condition for correct measurements of conductivity is linear distribution of potential across the specimen; this condition was satisfied in our measurements.

In the light of these considerations, our measurements were performed by Townsend's method at $\rho > 10^{12}$ ohms. As the time required for establishment of polarization at low temperatures is many hours, the temperature of the specimen was lowered under applied potential in order to accelerate polarization. The measurement times at low temperatures were decreased considerably in this manner.

It was also found that surface conductivity can introduce large errors at temperatures considerably above 100°C (similar results were obtained by Z. A. Levtsova [3]). The triangles in Fig. 3 correspond to results obtained for glass 90K without steps being taken to prevent surface conduction. Reproducible results could be obtained only when the specimens had been treated with water-repellent agents.

The glasses chosen for the determinations were those for which data are available in the literature on the resistance–temperature relation. The compositions of the glasses are given in the table. The results are plotted in Fig. 3. It follows from Fig. 3 that the dependence of resistance on temperature for the glasses studied conforms to Eq. (1) up to 10^{17} ohm-cm.

Literature Cited

1. O. V. Mazurin, "Electrical properties of glass," Tr. Leningr. Tekhnol. Inst. im. Lensoveta (62) (1962). [A partial translation of this book has been included as an introduction to this collection.]
2. L. Yu. Kurtts, Izv. Akad. Nauk SSSR, Otd. Khim. Nauk (5) (1940).
3. K. S. Evstrop'ev, this collection, p. 59.
4. N. M. Verebeichik and V. I. Odelevskii, Collection: The Glassy State, Izd. Akad. Nauk SSSR (1960) p. 282. [English translation: The Structure of Glass, Vol. 2, Consultants Bureau, New York (1960) p. 248.]
5. V. A. Ioffe and G. I. Khvostenko, Fiz. Tverd. Tela, 2:509 (1960).
6. I. I. Kitaigorodskii (ed.), Glass Technology, Gosstroiizdat (1961) p. 550.

TRANSPORT OF POSITIVE SODIUM AND POTASSIUM IONS IN A DIRECT ELECTRIC FIELD AT MOLTEN SALT–ALKALI GLASS–VACUUM INTERFACES

V. I. Danilkin, L. A. Kudryavtsev, and V. A. Ivanov

Experimental

Test tubes 250 mm long, 20-35 mm in diameter, and 1-1.5 mm thick were made from the test glasses. The inner surfaces of some of the specimens were provided with electrodes to ensure electrical contact. Platinum cathodes were applied by cathodic sputtering; molybdenum cathodes by thermal sputtering in a high vacuum; conducting layers of a colloidal graphite suspension were also used. The specimens were degassed for a long time before the determinations and then immersed in molten sodium or potassium nitrate in a cylindrical nickel bath. The glass was then electrolyzed for a long time at temperatures and current densities close to the experimental values. The apparatus is shown diagrammatically in Fig. 1. The vacuum system to which the glass specimen was welded consisted of an RVN-20 mechanical pump, a mercury vapor diffusion pump, and a system of liquid nitrogen freezing traps. The vacuum was measured by means of the VIT-1A vacuum meter and LM-2 and LT-2 tubes. The temperature of the bath with the molten electrolyte was maintained constant. A stabilized voltage was applied to the carbon anode and the internal coating. The alkali-metal ions were neutralized at the internal coating of the glass. The alkali-metal atoms evaporated from the cathode surface and were deposited on the cooled surface of the collector. The collector was a nickel cylinder fitted over a glass tube filled with liquid nitrogen. At the end of an experiment the system was filled with argon; the nickel

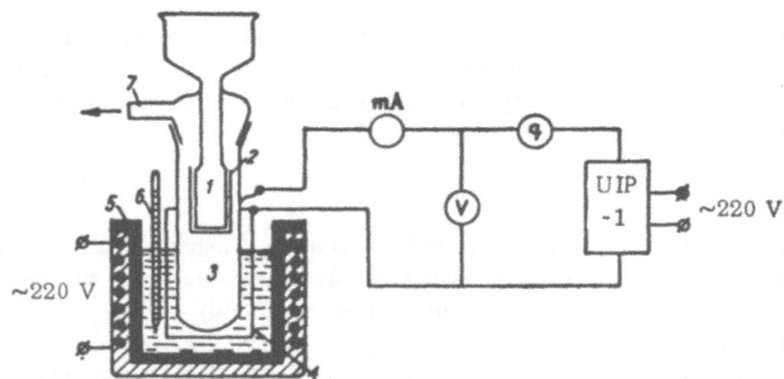

Fig. 1. Electrical circuit for electrolysis of alkali glass specimens. 1) Trap with liquid nitrogen; 2) nickel collector; 3) glass specimen with conductive coating; 4) carbon electrode; 5) nickel bath with molten alkali-metal nitrate; 6) mercury thermometer with extended scale; 7) tube to vacuum system; mA) milliammeter; V) voltmeter; q) Lingane gas coulometer; UIP-1) universal current source.

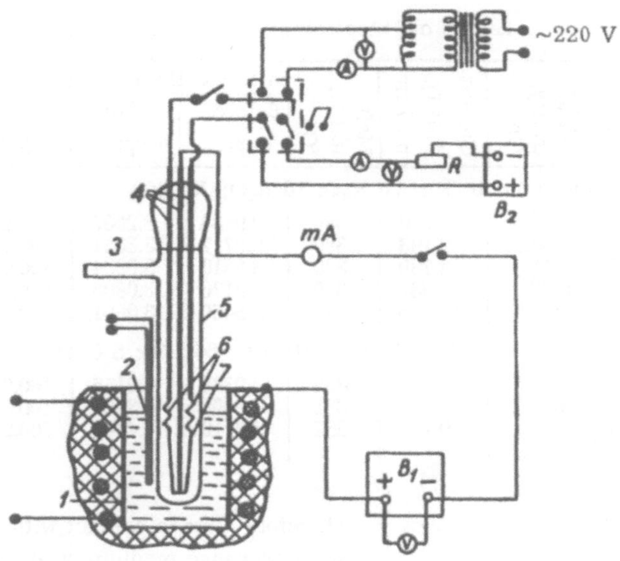

Fig. 2. Line diagram of the experimental unit. 1) Bath with molten electrolyte; 2) thermocouple; 3) tube to vacuum system; 4) molybdenum conductors; 5) glass specimen, 1.5 mm thick; 6) springs; 7) tungsten filament; B_1) UIP-1 rectifier; B_2) VSA-6M rectifier; V) voltmeter; A) ammeter; mA) milliammeter.

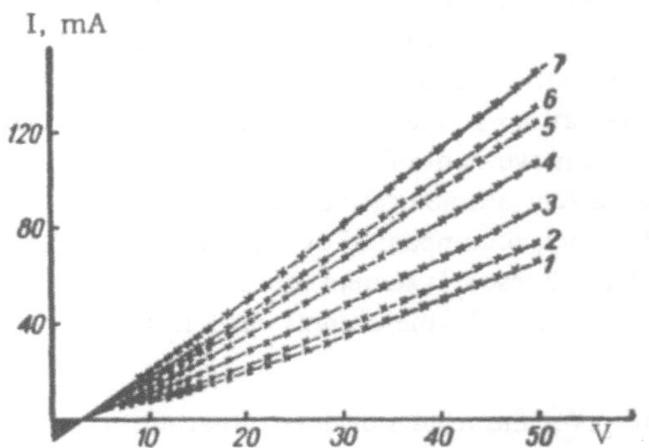

Fig. 3. Typical volt–ampere characteristics for transport of potassium through potassium tetrasilicate glass with an internal platinum coating at different temperatures.

collector with the deposit of alkali metal was put into a glass test tube, also filled with argon, and weighed on an analytical balance to the nearest 0.1 mg. The quantity of electricity passed through the glass specimen was determined with a Lingane coulometer [1] containing 0.5 N K_2SO_4 as the working solution. The precision of the coulometer was 0.1%. The vacuum in the evacuated space was $2.0 \cdot 10^{-5}$ mm Hg. Figure 2 is a diagram of the apparatus for investigating glasses without metallic coatings. The alkali-metal ions liberated at the glass surface were neutralized by thermoelectrons. The electron emitter was a U-shaped tungsten filament 100 μ in diameter and 150 mm long. The filament was heated by direct current. Use of alternating current for heating the filament did not alter the experimental results. The temperature of the emitter was estimated from the power consumption required to heat it, with a correction for cooling of the filament ends by the holders [2], and also by means of the OPIR-17 optical pyrometer. The central point of the emitter was maintained at a negative potential from a stabilized voltage source. Instruments of the 0.2 class were used for the current and voltage measurements. The vacuum in the evacuated space was $2 \cdot 10^{-6}$ mm Hg.

Results and Discussion

If the alkali glass under investigation is in a molten salt with a common exchange ion which is the main carrier of current through the glass, and an insoluble electrode is used as the anode, destruction of the glass never occurs during prolonged electrolysis in a direct field. If an anode with a different exchange ion is used, the glass is attacked by introduction of "foreign" ions into its structural elements. Positive potassium ions, penetrating into soda glass, change its structure under the influence of the direct field. Such glass cracked on cooling.

Results of "Electrolysis" of Potash Glasses

No.	Current, mA	Quantity of electricity, Cb	Current density, mA/cm^2	Melt temp., °C	Wt. of collector, g: Before expt.	Wt. of collector, g: After expt.	Current efficiency, %	$E_{expt} \cdot 10^4$, g/Cb
Glass 1 (mole %): 20 K_2O; 10 B_2O; 70 SiO_2								
1	50	536.5	1.096	380	2.9833	3.2007	99.9	4.0521
2	100	896.82	2.193	380	2.9717	3.3351	99.6	4.0521
3	200	1111.5	4.789	380	3.5310	3.0804	100.0	4.0539
4	350	1078.2	8.381	380	3.5176	3.0805	99.9	4.0539
5	450	1149.7	10.775	380	3.5460	3.0801	99.9	4.0523
Glass 2 (mole %): 20 K_2O; 10 B_2O; 65 SiO_2; 5 CaO								
1	50	691.3	1.52	382	3.9343	4.2156	100.0	4.0561
2	150	961.2	4.55	381	4.0098	4.3987	100.0	4.0522
3	300	1143.1	9.1	382	4.0110	4.4742	100.0	4.0521

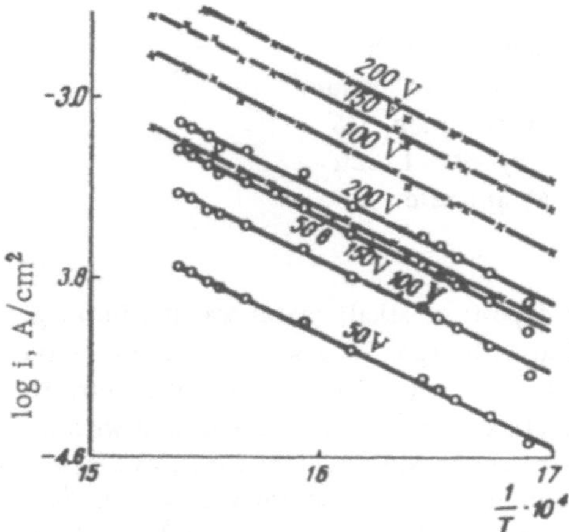

Fig. 4. Plots of the logarithm of the current density in glasses with internal molybdenum coatings versus the reciprocal temperature of the melt. O) Glass No. 46; ×) sodium borosilicate glass.

Therefore glass in contact with the melt should be regarded as an exchange medium or as a special case of an electrolyte solution with a low dielectric constant [3].

Let us consider the concrete case of transport of positive sodium and potassium ions through alkali glasses into vacuum, with the subsequent neutralization of the ions (Fig. 3). In a direct field, positive ions jump from the alkali-metal nitrate melt into the vacancies in the glass, which originate mainly by electrolytic dissociation of the cations which leave the polar structural elements and pass into the interstitial space of the network [3]. The concentration of vacancies in the glass remains constant at constant temperature. Directed movement of the cations in the direction of the field toward the cathode causes movement of the vacancies in the opposite direction. The cations are neutralized at the glass–cathode boundary. This is not accompanied by breakdown of polar groups in the glass. Faraday's second law is applied for determination of the charge of the current carriers through alkali glasses.

The table gives the results of determinations of the reciprocal specific charge of the carriers of current through potassium glass, and the corresponding current efficiencies. It is seen that, within the limits of experimental error, the amount of alkali metal discharged corresponds to the quantity of electricity passed. In this case the electricity is completely consumed for ionic transport of potassium or sodium from melts of the respective nitrates through the glass into vacuum, followed by neutralization of the ions.

If we plot the logarithm of the current density i through the glass against the reciprocal absolute temperature (Fig. 4), we obtain a series of parallel straight lines the slope of which gives the effective activation energy ΔW.* The value of ΔW for the soda glasses studied is of the order of 0.9 eV. It is interesting to note that the values found by the usual method of conductivity determination for the total activation energies of a number of specimens [4] differ from those obtained by the present method. Moreover, plots of log current density versus the reciprocal temperature for specimens of the composition (mole %): $20K_2O$, $80SiO_2$, and $20K_2O$, $10B_2O_3$, $70SiO_2$ show distinct breaks (two linear regions of different slopes). For the borosilicate glass ΔW found from the first linear region, in the temperature range from 336 to 360°C, is 1.1 eV, while the value from the second

*From the formula $i = A \exp(-\Delta W/2kT)$, where A is a constant and k is the Boltzmann constant.

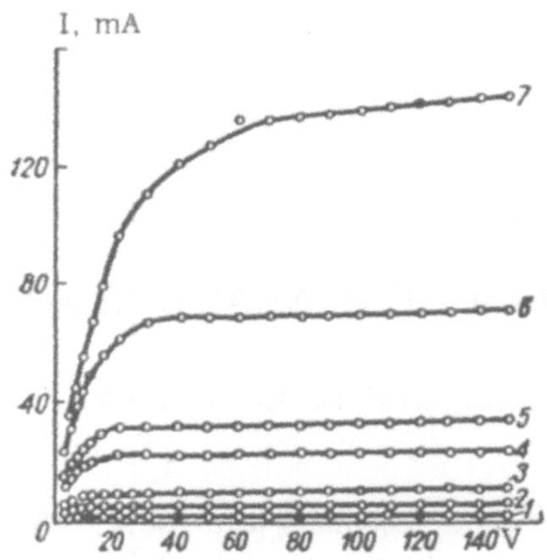

Fig. 5. Volt–ampere characteristics of potassium tetrasilicate glass at different temperatures of the electron emitter (°C): 1) 2000; 2) 2050; 3) 2100; 4) 2140; 5) 2180; 6) 2220; 7) 2270. Melt temperature 340°C.

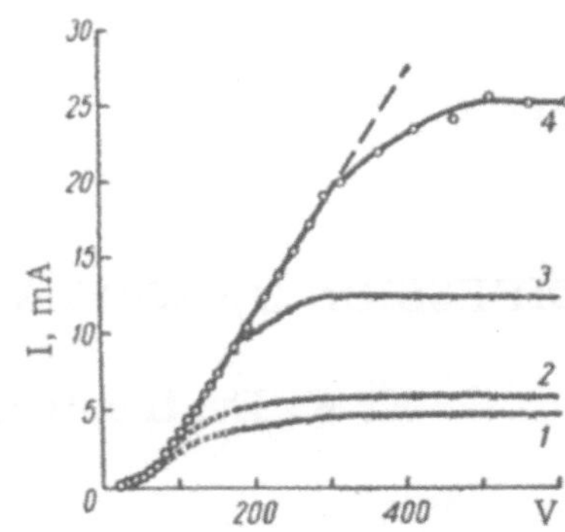

Fig. 6. Variations of the ionic current through glass No. 46, without an internal electrode, with the applied voltage at different temperatures of the electron emitter (°C): 1) 2080; 2) 2125; 3) 2180; 4) 2232. Temperature of salt melt 380°C.

region, from 360 to 400°C, is 0.8 eV. For potassium tetrasilicate, ΔW is 0.48 eV in the first region, up to 380°C, and 0.22 eV in the second region, from 380 to 400°C. The causes of these breaks have not been established. However, they can hardly be attributed to inaccuracies in the method. In this temperature range soda glasses give linear plots of log i versus 1/T (without breaks). The main cause of the breaks is apparently associated with structural characteristics of the glass.

If the positive sodium ions are neutralized at the glass surface by electrons from a tungsten emitter, the current flowing through the glass is evidently found from the equation

$$I = gV^{3/2}$$

where V is the potential, g is a coefficient which depends on the configuration and distance between the electrodes, and also on the specific charge of the current carriers. The volt–ampere characteristics of this process, determined at constant melt temperature and different temperatures of the tungsten emitter, are given in Figs. 5 and 6.

Literature Cited

1. S. S. Lingane, J. Am. Chem. Soc., 67:1916 (1945).
2. B. M. Tsarev, Electronic Tube Design Calculations, Gosénergoizdat (1961).
3. R. L. Myuller, Zh. Fiz. Khim., 6:616 (1935); Zh. Fiz. Khim., 25:236, 246 (1955).
4. O. V. Mazurin, "Electrical properties of glass," Tr. Leningr. Tekhnol. Inst. im. Lensoveta (62) (1962). [A partial translation of this book has been included as an introduction to this collection.]

DIFFUSION OF SODIUM IONS IN SODIUM GERMANATE GLASSES

K. K. Evstrop'ev, V. K. Pavlovskii, and A. O. Ivanov

The glasses were made from Na_2CO_3 of "pure grade" and "spectroscopically pure" GeO_2. The diffusion coefficients were determined by the technique developed earlier [1], with the use of the Na^{22} radioisotope as the tracer.

The experimental data are presented in Fig. 1. Conductivities of the glasses as given in [2] are shown on the same diagram. The divergence between the D_{Na} and $\varkappa$ curves in the low-alkali region can be explained with the aid of the Einstein equation:

$$\varkappa = \frac{DNe^2}{kT}$$

where $\varkappa$ is the conductivity, D is the coefficient of diffusion, N is the number of current carriers per unit volume, e is the electronic charge, k is the Boltzmann constant, and T is the absolute temperature.

It is seen that variations of $\varkappa$ are determined by superposition of two factors: variations in the volume concentration of the current carriers, and changes of their mobility The first additions of Na_2O raise the volume concentration of Na ions so rapidly that the conductivity increases despite the decrease of D_{Na}. In glasses with over 10 mole % Na_2O the variations of conductivity are determined mainly by changes of D_{Na}. Therefore in this concentration range the plots of log D_{Na} and log $\varkappa$ versus the concentration follow a similar course.

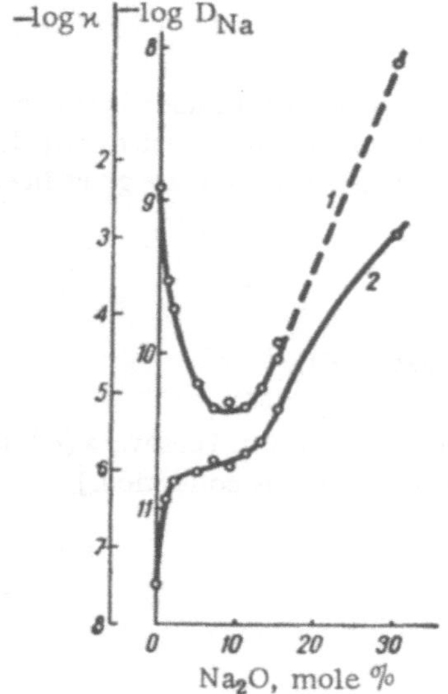

Fig. 1. Variations of diffusion (1) and conductivity (2) with glass composition at 415°C.

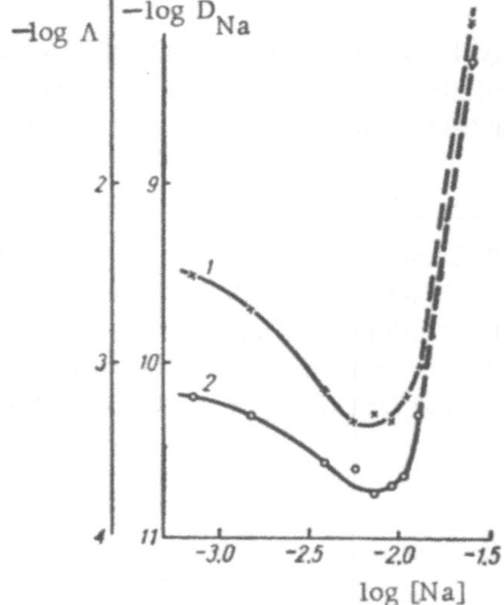

Fig. 2. Variations of diffusion (1) and equivalent conductivity (2) with the volume concentration of alkali-metal ions in the glass at 415°C.

The equivalent conductivity Λ, calculated per mole of the alkali component, depends on D_{Na} only. Therefore the variations of log D_{Na} and log Λ with the ion concentration in the glass should be entirely similar; this is confirmed by Fig. 2.

Figure 2 shows that over a certain concentration range D_{Na} and Λ decrease with increasing Na_2O content in the glass. The existence of a similar relationship between log Λ and log [Na] for borate glasses suggests that, as in borate glasses [3], the increase of dissociation energy and the associated decrease of equivalent conductivity of germanate glasses are due to increase of the Madelung constant. Indeed, at low alkali concentrations the dissociation energy is determined by the work function of the cation passing from the polar group into the nonpolar medium.

The conductivities of the glasses were calculated from the experimental diffusion coefficients with the aid of the Einstein equation. These calculated values agree, within half an order of magnitude, with the experimental conductivities; this confirms the hypothesis of ionic conductivity of germanate glasses. Calculations show that glassy germanium dioxide is also an ionic conductor.

Literature Cited

1. K. K. Evstrop'ev, Author's summary of candidate's dissertation, Leningr. Gos. Univ. im. A. A. Zhdanova (1962).
2. A. O. Ivanov, K. S. Evstrop'ev, and M. L. Dorokhova, Optiko-Mekh. Prom. (1):31 (1962).
3. R. L. Myuller, Zh. Tekhn. Fiz., 26:2614 (1956).

ELECTROCHEMICAL METHOD FOR DETERMINING DIFFUSION COEFFICIENTS OF SILICATE MELTS

N. K. Dertev and Z. P. Voronkova

Diffusion processes in silicate melts were studied with the aid of concentration cells with platinum electrodes, of the type

$$\mathrm{Pt} \mid \text{melt I} \mid \text{melt II} \mid \mathrm{Pt}$$

In the course of diffusion the potential difference varies with the difference of concentrations in the melt. Determinations of the changes of potential difference with time at constant temperature therefore give an indication of the diffusion rate and provide the necessary data for determination of diffusion coefficients.

This method for determination of diffusion coefficients, first used by A. F. Borisov [1], we describe as the electrochemical method. We used this method for determining the interdiffusion coefficients of silicate melts. A new technique was developed for this purpose.

The cell is shown schematically in Fig. 1. The potential difference was determined by the balancing method with the aid of the PPTV-1 potentiometer. Experimental calibration curves obtained by means of preliminary experiments were used for converting potential differences to concentration gradients.

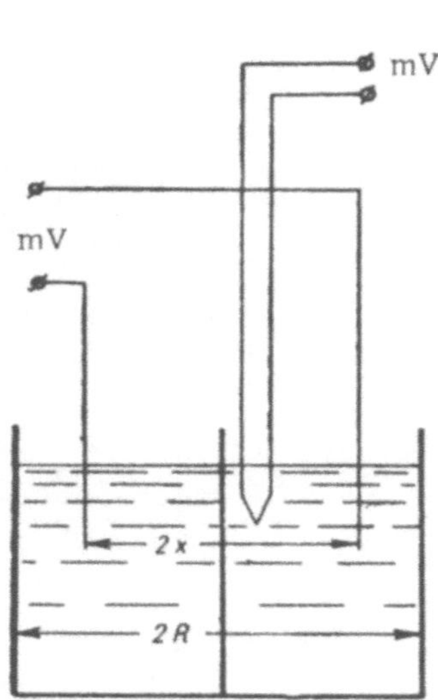

Fig. 1. Diagram of cell for determination of interdiffusion coefficients.

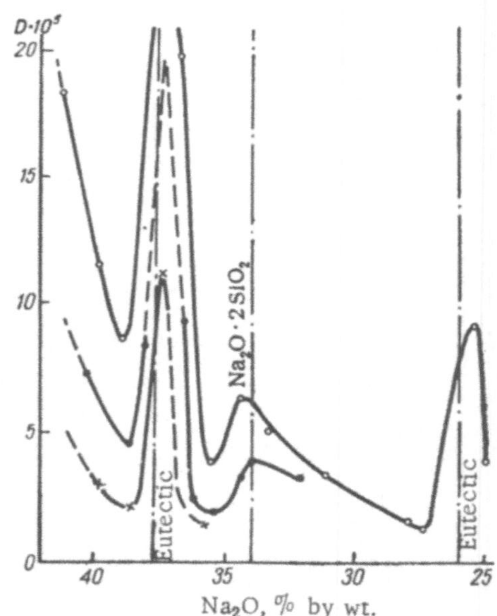

Fig. 2. Variations of the diffusion coefficients of sodium ions with composition of glasses in the system Na_2O-SiO_2. O) At 1100°C; ●) at 1000°C; ×) at 950°C.

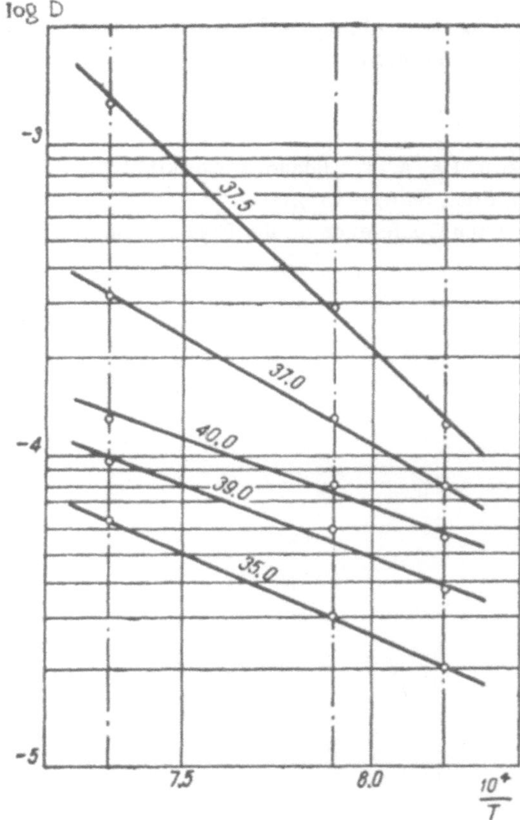

Fig. 3. Dependence of the diffusion coefficients of glasses in the system Na_2O-SiO_2 on temperature (the numbers on the lines correspond to Na_2O contents in % by weight).

The general diffusion equation

$$\frac{dC}{dt} = D\frac{d^2C}{dx^2} \tag{1}$$

where D is the diffusion coefficient and C is the concentration, was used for the calculations.

This equation can be used with small initial concentration differences. With equal distances x from the electrodes to the partition and equal cell volumes on the two sides of the partition, solution of this equation gives the expression

$$D = \frac{\ln P - \ln a}{b} \tag{2}$$

where

$$a = \frac{4}{\pi} \cdot \sin\frac{\pi x}{2R}; \quad b = -\left(\frac{\pi}{2}\right)^2 \frac{\tau}{R^2}; \quad P = 2\left(\frac{C}{C_0} - \frac{1}{2}\right)$$

Here R is the cell length (distance from the partition to the opposite wall of the crucible, equal for the two cells).

We used this method for determining the interdiffusion coefficients of glasses in the system $Na_2O-CaO-SiO_2$. The fullest data were obtained for the system Na_2O-SiO_2.

The diffusion coefficients so found were referred to the melt formed after completion of the diffusion process.

The diffusion coefficients are shown in Fig. 2. It is seen in Fig. 2 that the relationship between the diffusion coefficient and the composition corresponds completely to the equilibrium diagram of the system. In particular, the maxima correspond to eutectic melts; we attribute this to the special state of the eutectic.

To verify the results, they were evaluated with the aid of the equation for the diffusion coefficients as a function of the temperature:

$$D = D_0 \cdot e^{-\frac{U}{RT}} \tag{3}$$

whence

$$\log D = \log D_0 - \frac{U}{RT} \cdot \log e \tag{4}$$

The relation (4) is plotted in Fig. 3. We have only three experimental points (for 950, 1000, and 1100°C) for each melt, but the fact that they all lie exactly on straight lines for all the melts and therefore obey Eq. (4) confirms the validity of the results.

The error in calculation of diffusion coefficients from the results obtained by the electrochemical method is ±10%.

The literature contains only a few publications dealing with determinations of diffusion coefficients in silicate melts. They include the paper by Johnson, Bristow, and Blau [2], who determined diffusion coefficients with the aid of radioactive tracers. Their values for the diffusion coefficients of sodium oxide in a glass of the composition 34% Na_2O, 66% SiO_2 at 1000 and 1100°C agree with ours.

Malkin and Mogutnov [3] used a radioactive isotope method for determining coefficients of self-diffusion of sodium ions into silicate melts. Their values also agree with ours in order of magnitude.

Literature Cited

1. A. F. Borisov and N. K. Dertev, Tr. Gor'kovsk. Politekhn. Inst. im. A. A. Zhdanova, Gor'kii, 13(5) (1957).
2. J. R. Johnson, R. H. Bristow, and H. H. Blau, J. Am. Ceram. Soc., 34(6) (1951).
3. V. I. Malkin and B. M. Mogutnov, Proceedings of the All-Union Conference on the Physical Chemistry of Molten Salts and Slags (1962).

DIELECTRIC LOSSES OF GLASSES AT THE WAVELENGTH OF 8 MM*

M. D. Mashkovich

The values of ε and tan δ for a number of glasses were determined in the temperature range of 20-400°C at a wavelength of 8 mm with the aid of a cylindrical volume resonator. The results were compared with analogous data for the wavelength of 3.2 cm. It was found that "smoothing" of the tan δ versus t curves with increasing frequency, observed earlier † in a range of longer waves, did not occur in any instance.

Glasses containing Pb and K ions, with the lowest fundamental vibration frequency, show an appreciable intensification of the effect of temperature on tan δ at 8 mm. Glasses of these compositions also exhibit the greatest increase of tan δ with increasing frequency at superhigh frequencies. It is suggested that at room temperature the absorption spectra of glasses have a form characteristic for relaxation owing to elastic ionic vibrations. The intensified effect of temperature on tan δ at 8 mm in comparison with 3.2 cm is attributed to broadening of the absorption curve with rise of temperature.

* Paper published in Fiz. Tverd. Tela 5 (6): 1740 (1963). [English transl. Soviet Physics - Solid State 5 (6): 1265 (1963) Consultants Bureau, New York.]

† M. D. Mashkovich, Fiz. Tverd. Tela 3 (4): 1105 (1961). [English transl. Soviet Physics - Solid State 3 (4): 804 (1961) Consultants Bureau, New York.]

CHANGES IN THE ELECTRICAL PROPERTIES OF GLASSES IN THE SYSTEM $MgO-Al_2O_3-SiO_2$ DURING CRYSTALLIZATION

G. A. Pavlova

The resistivities and dielectric losses of original and crystallized glasses in the system $MgO-Al_2O_3-SiO_2$ were studied at elevated temperatures.

This system is the basis of crystalline glass materials which are of practical interest [1, 2].

The specimens were crystallized under the same conditions, being held for six hours at 1000-1200°C and for two hours at 1200°C.

The results of the resistivity determinations are shown in Figs. 1 and 2.

Results for the original glasses (Fig. 1) show that replacement of SiO_2 by MgO or Al_2O_3 raises the resistance, and that the effect of MgO is somewhat greater than that of Al_2O_3.

The effect of composition on variations of dielectric losses with temperature for these glasses is analogous to the effect of composition on the resistivity variations. The relation found between the electrical properties and the composition of the glasses can, in our opinion, be attributed to impurity conduction in these glasses.

Crystallization lowers the resistivity and activation energy of conduction of the glasses. The higher the resistance of the original glass and, hence, the lower its SiO_2 content, the less is the decrease after crystallization. The activation energy of conduction decreases by 20-40% in crystallized glasses.

The effect of composition on the resistance of crystallized glasses at 500°C is illustrated by Fig. 2. In contrast to the original glasses, in this case Al_2O_3 has a greater influence than MgO on the increase of resistance.

Dielectric losses at elevated temperatures and 10^6 cps are much greater in was crystallized than in the original glasses.

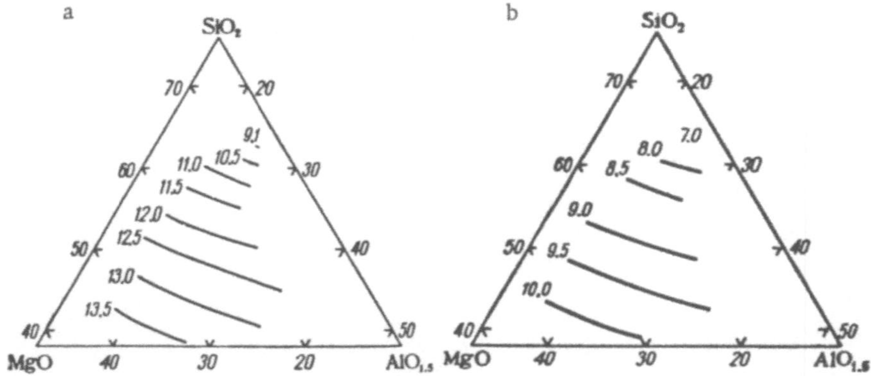

Fig. 1. Lines of equal log resistance ("isoresists") for the original glasses (compositions in cation %). a) At 300°C; b) at 500°C.

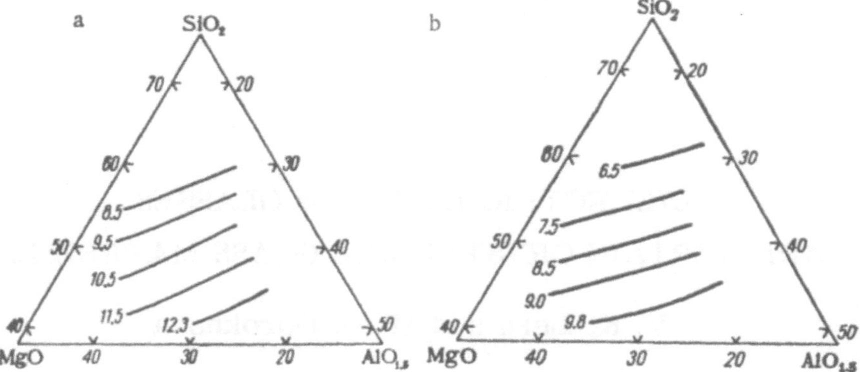

Fig. 2. Lines of equal log resistance ("isoresists") for crystallized glasses. a) At 300°C; b) at 500°C.

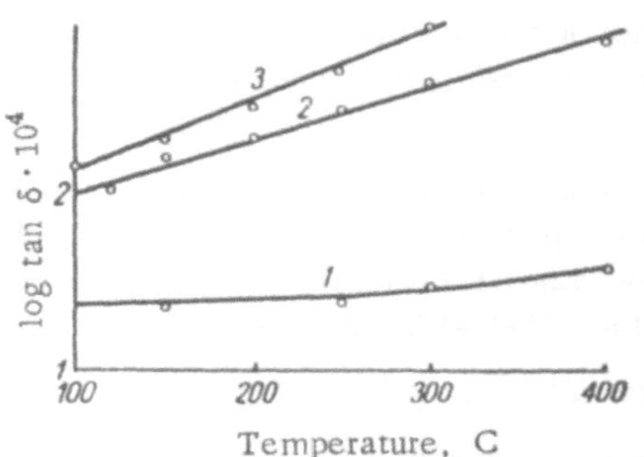

Fig. 3. Effect of temperature on tan δ at 10^6 cps for glass of the composition (% by wt.): 20MgO; $30Al_2O_3$; $50SiO_2$. 1) Original glass; 2) glass crystallized at 1050°C during 8 hr; 3) glass crystallized by the procedure used for all the other glasses.

The increased losses in crystallized glasses at temperatures up to 500°C are due mainly to the relaxation component, while conduction losses remain lower by several orders of magnitude. As an example, Fig. 3 shows data on the tan δ versus temperature relationship for one of the glasses and for specimens of the same glass crystallized under different conditions. It is seen in Fig. 3 that, despite the high resistivity of the crystallized specimens, their dielectric losses are considerably higher than those of the original glass.

The composition has very much less effect on the relationship between dielectric losses and temperature for crystallized glasses than on the corresponding relationship for resistivity. The maximum change of tan δ with the composition at 10^6 cps is about 0.5 of an order of magnitude.

The results suggest that in crystallized glasses a considerable increase of the relaxation loss component is due to separation of the crystalline phase and to the structure of the material as a whole. We consider that the dielectric losses of crystallized glasses are predominantly influenced by α-cordierite, the content of which, according to the results of x-ray diffraction determinations, increases with decrease of the SiO_2 and increase of the Al_2O_3 contents. Impurities, interphase boundaries, and cracks increase these losses. The resistivity is determined by the contacting interlayers of residual glass, which has the higher resistance. These data indicate that with crystallized glasses in the system $MgO-Al_2O_3-SiO_2$ there is no analogy between the effects of composition on resistivity and dielectric losses.

In the case of lithium aluminosilicate glasses the resistivity and activation energy of conduction increase during crystallization [3]; in contrast to this, the resistivity and activation energy of magnesium aluminosilicate glasses decrease during crystallization. Increase of Al_2O_3 concentration lowers the resistivity of crystallized lithium silicate glasses, whereas in crystallized glasses of the cordierite system the resistivity increases with increasing Al_2O_3 concentration.

Literature Cited

1. J. E. Burke, Progress in Ceramic Science, Vol. II (1962).
2. Collection: The Glassy State, Vol. III, No. 1, Catalyzed Crystallization of Glass, Izd. Akad. Nauk SSSR (1963). [English translation: The Structure of Glass, Vol. 3, Consultants Bureau, New York (1964).]
3. G. A. Pavlova and V. G. Chistoserdov, Collection: The Glassy State, Vol. III, No. 1, Catalyzed Crystallization of Glass, Izd. Akad. Nauk SSSR (1963) p. 184. [English translation: The Structure of Glass, Vol. 3, Consultants Bureau, New York (1964) p. 200.]

DIELECTRIC LOSSES IN GLASSES AND CERTAIN CRYSTALLINE GLASS MATERIALS

V. K. Leko and M. L. Dorokhova

The microheterogeneous structure of glass is now fully proved. According to Maxwell [1] and Wagner [2], heterogeneous structure in dielectrics invariably leads to electrical aftereffects, resulting in a decrease of current with time in a direct field, and causing dielectric losses in an alternating field.

We base our explanation of the origin of dielectric losses in glasses on the concepts of R. L. Myuller [3] concerning the nature of conduction in simple alkali glasses. These concepts can probably be extended to some degree to systems of greater complexity.

The presence of conducting polar groups in a poorly conducting nonpolar medium gives rise to relaxation processes when an electric field is applied, with relaxation times inversely proportional to the conductivity of the polar microinclusions. The diversity of the forms, sizes, and relative amounts of the polar groups leads to the appearance of a wide distribution of relaxation times, which should not alter with temperature [4].

Migration losses in glasses consist of relaxation losses and conduction losses.

In the case of low-alkali glasses, where conduction losses are due to movement of the ions in the nonpolar medium with a relatively high activation energy, while the relaxation losses are determined by movement of the ions within the enclosed polar groups with a lower activation energy, the two types of losses can be easily separated by variations of temperature and frequency.

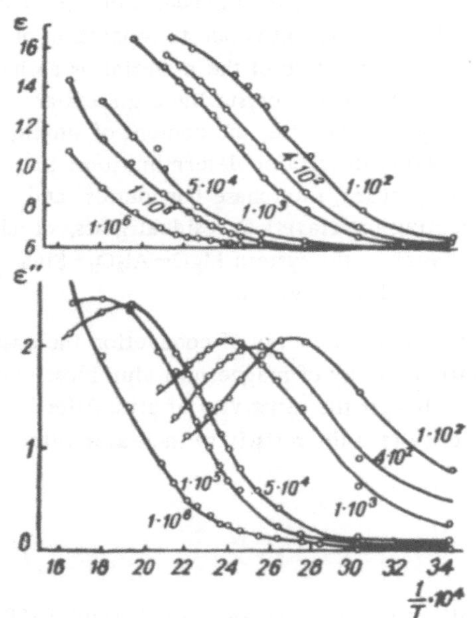

Fig. 1. Temperature–frequency relations of the loss factor and dielectric constant of a glass of the composition (mole %): $5Na_2O$, $95SiO_2$.

Our measurement of dielectric losses in a glass of the composition 5% Na_2O, 95% SiO_2 (Fig. 1) reveals clear relaxation loss maxima.

The activation energy of diffusion of sodium ions within the enclosed polar groups, calculated from the displacement of the maximum, is 0.6 eV.

The activation energy of sodium-ion diffusion in the nonpolar medium, found from conductivity data, is 1.1 eV. The low activation energy of ion movement within the closed groups (approximately the same as for a glass of the composition 30% Na_2O, 70% SiO_2) indicates a high concentration of sodium ions in these regions.

Such relations were not found for glass of the composition 5% K_2O, 95% SiO_2.

If relaxation losses and conduction losses are of the same nature (caused by migration of ions along interconnected polar regions), their activation energies are equal. In that case, as has been shown by Isard [5] and Mazurin [6], the migration losses plotted against log $f\rho$ lie on smooth curves.

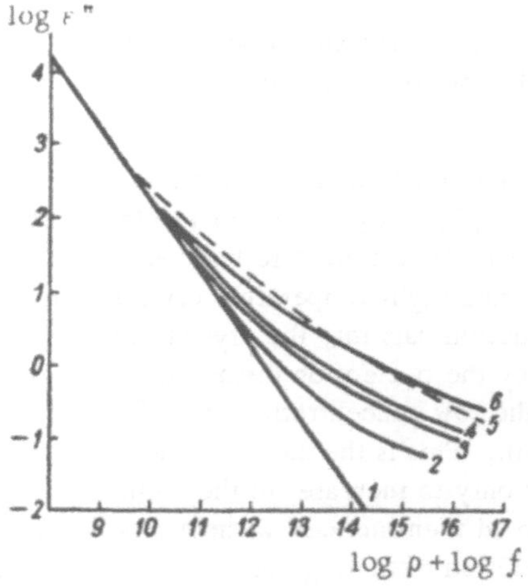

Fig. 2. General character of the dependence of the loss factor on frequency and resistivity for various glasses: 1) Conductivity line; 2) alkali-free glasses; 3) photosensitive glass held at 600°C for 1 hr; 4) high-alkali glasses; 5) glass with electronic conduction; 6) photosensitive glass held at 800°C for 1 hr.

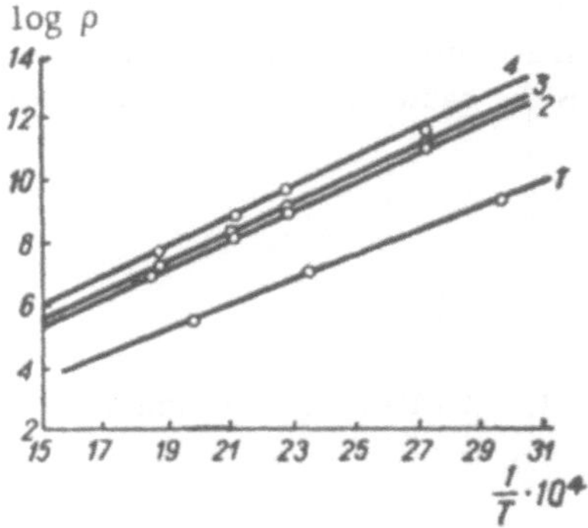

Fig. 3. Effect of temperature on resistivity of glass materials before and after crystallization: 1) Original photosensitive glass; 2) glass crystallized for 1 hr at 600°C; 3) glass crystallized for 1 hr at 700°C; 4) glass crystallized for 1 hr at 800°C.

The distribution range of relaxation times can be qualitatively assessed from the position of these curves in relation to the conduction loss curve.

If dielectric losses in glasses are caused by their microheterogeneous structure, the nature of the losses should be the same for glasses with different types of conduction. To verify this hypothesis, we determined losses in three types of glasses with different types of conduction:

1. High-alkali glasses: $Li_2O \cdot 2.75SiO_2$; $3Na_2O \cdot 2Al_2O_3 \cdot 5B_2O_3$; $70.8SiO_2$, $4.4Al_2O_3$, $22.4Li_2O$, $2.4K_2O$ (in mole %).
2. Alkali-free glasses of the series (mole %): 50RO, $50SiO_2$, where RO is CaO, BaO, or PbO.
3. Alkali-free glass containing about 5% Ti^{3+} and having electronic conduction.

Data on the dielectric losses of all these glasses fit on smooth curves in plots of log ε" versus log ρf (Fig. 2).

The dielectric losses of alkali glasses are characterized by curve 4, the empirical equation for which was given by Mazurin [6]:

$$\log \varepsilon'' = -1.9 + 40 \cdot 10^{-0.1 \log \rho f}$$

This indicates that all high-alkali glasses have almost the same distribution of relaxation times, thus showing a similar distribution of heterogeneities in them.

Alkali-free glasses are characterized by curve 2. Their distribution of relaxation times is narrower than for alkali glasses.

The electronically conducting glass has the broadest distribution of relaxation times (curve 5).

Thus, the nature of the conduction influences only the breadth of the distribution of relaxation times, while the remaining relations found for alkali glasses are valid for glasses having other types of conduction.

This type of loss, due to microheterogeneity of the glass, should apply to any solid dielectric containing heterogeneity regions having considerably higher conductivity than the surrounding medium. An example of such a material is the completely crystallized photosensitive glass of the composition (mole %): $70.8SiO_2$, $4.4Al_2O_3$, $22.4Li_2O$, $2.4K_2O$.

Study of the structure of this crystalline glass material revealed the presence of isolated groups of lithium silicates and aluminosilicates surrounded by interlayers close to cristobalite in composition [7]. Photomicrographs show that some regions enriched with lithium silicates and aluminosilicates are in contact with each other and form continuous threads.

This structure leads to the conclusion that relaxation losses and conduction losses are of a common nature.

It is interesting to note that heat treatment of crystalline glass material broadens the distribution of relaxation times by changing its structure (Fig. 2).

In Fig. 3, log ρ of the material is plotted against the temperature, before and after crystallization. The increase of log ρ_0 by an order of magnitude indicates that only about 10% of all the Li^+ ions is involved in through conduction.

Another example is provided by a crystalline glass material in the cordierite region with an admixture of Ti^{3+}. It is known that crystalline glass materials of the system $MgO-Al_2O_3-SiO_2-TiO_2$, containing small amounts of trivalent titanium, exhibit distinct relaxation maxima on the plots of dielectric loss versus the temperature and frequency [8]. This effect can be explained as follows. During high-temperature crystallization of these materials trivalent titanium compounds having electronic conduction pass into the crystalline phase. The activation energy of conduction of these compounds is determined by the p-electron dissociation energy $Ti^{3+} \rightarrow Ti^{4+} + e$, which is approximately 0.5 eV ($E_A + \frac{1}{2}E_d$). Because of the low concentration of Ti^{3+}, the electronically conducting phases are dispersed in a poorly conducting medium. This is the cause of the relaxation losses. Increase of the Ti^{3+} concentration to a certain limit should lead only to increase in the number and dimensions of the conducting regions. The dielectric loss maximum should then increase without any significant change in the activation energy, equal to the activation energy of conduction of the regions of high conductivity. On further increase of the Ti^{3+} concentration these regions begin to make mutual contact, ensuring through conduction. The distinct loss maximum should then no longer be observed.

To confirm this hypothesis, we made a number of glasses with high contents (~5%) of reduced titanium. These glasses are electronic conductors with activation energy in the region of 1 eV. The activation energy of conduction was altered very little when the glasses were crystallized at temperatures below 1000°C. This leads to the conclusion that after crystallization Ti^{3+} remains in the glassy phase, which limits the conductivity of the specimens. After crystallization at temperatures above 1000°C the activation energy fell to 0.3 eV, apparently because Ti^{3+} passed into the crystalline phase. Such materials do not exhibit evident relaxation dielectric loss maxima.

It should be noted that at about 1000°C cordierite begins to crystallize in glasses of the system $MgO-Al_2O_3-SiO_2-TiO_2$. Because of this, cordierite was erroneously regarded as the cause of the relaxation process, to which an ionic character was attributed [8].

Literature Cited

1. J. C. Maxwell, Treatise on Electricity and Magnetism, Vol. 1, Berlin (1883).
2. C. Wagner, Arch. Electrotechn.,2:371 (1914).
3. R. L. Myuller, Uch. Zap. Leningr. Gos. Univ. im. A. A. Zhdanova (54):159 (1940); Zh. Tekhn. Fiz.,26:2614 (1956).
4. H. E. Taylor, Trans. Faraday Soc., 52:873 (1956).
5. J. O. Isard, Proc. Inst. Elec. Engrs. (London), B109, Suppl. No. 22 (1962).
6. O. V. Mazurin, "Electrical properties of glass," Tr. Leningr. Tekhnol. Inst. im. Lensoveta (62) (1962). [A partial translation of this book has been included as an introduction to this collection.]
7. N. A. Shmeleva, V. G Chistoserdov, and A. I. Gerasimova, Collection: The Glassy State, Vol. 3, No. 1, Catalyzed Crystallization of Glass, Izd. Akad. Nauk SSSR (1963) p. 159. [English translation: The Structure of Glass, Vol. 3, Consultants Bureau, New York (1964) p. 169.]
8. M. D. Mashkovich, Fiz. Tverd. Tela, 5:843 (1963).

DIELECTRIC LOSSES IN LEAD BORATE GLASSES

G. Kh. Kudashev

Dielectric losses of lead borate glasses at 50 cps were investigated.

The compositions of the glasses are given in Table 1. The results are plotted in Fig. 1.

Over a considerable temperature range the dielectric losses of the glasses (with the exception of glass No. 8) remain almost constant and very low.

At a certain temperature the losses begin to increase rapidly. The starting point depends on the lead oxide content of the glass: the loss rise begins at a lower temperature with increase of the lead oxide content. Three different positions can be occupied by the lead ion in borate glass (Fig. 2).

1. The lead ion may be between two oxygen atoms with unsaturated valences (ion A). Such ions strengthen the glass structure. They cannot be involved in transport of current and have virtually no effect on losses.
2. The lead ion may be attached to a single oxygen atom with unsaturated valence (ion B). This ion is bonded less strongly to the glass structure than ion A, but the bond is still fairly strong and the energy required to break it is relatively high. Therefore at low temperatures ions of type B have no appreciable influence on dielectric losses.
3. The lead ion is not bonded to any particular atom (ion C) and can easily move within the unit cell of the structural network or pass into neighboring cells. Directed displacements of such ions within the unit cell give rise mainly to relaxation losses, while directed jumps into neighboring cells result in conduction and in conduction losses in the glass at low temperatures.

The probability w_1 of directed displacement of an ion within the cell can be represented by the expression

$$w_1 = Ae^{-U/kT} \tag{1}$$

and the probability w_2 of directed transfer of the ion into a neighboring cell by the expression

$$w_2 = Ae^{-U'/kT} \tag{2}$$

In these equations U and U' are the energy barriers surmounted by the ion in the displacements, and $w_2 \ll w_1$.

TABLE 1

Glass No.	Comp., mole %		Glass No.	Comp., mole %	
	B_2O_3	PbO		B_2O_3	PbO
1	97.6	2.4	5	54.4	45.6
2	78.1	21.9	6	48.2	51.8
3	70.8	29.2	7	39.8	60.2
4	59.0	41.0	8	28.5	71.5

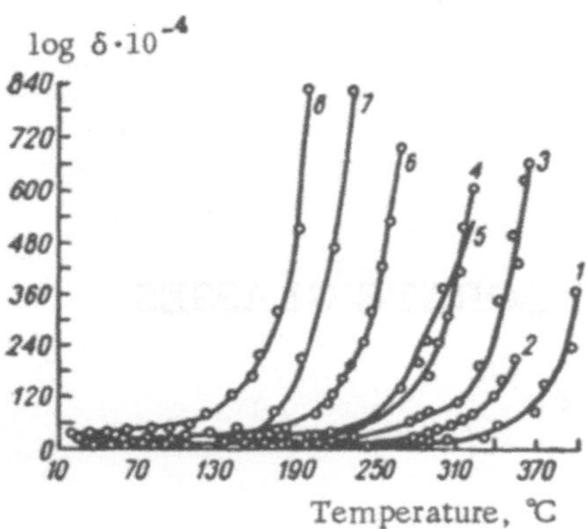

Fig. 1. Variations of the tangent of the loss angle with temperature for lead borate glasses.

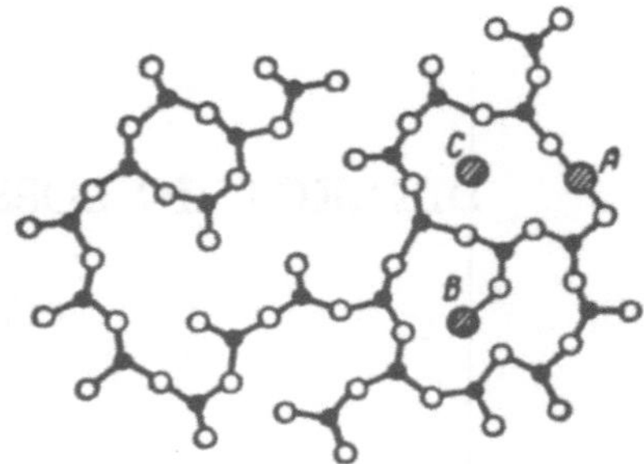

Fig. 2. Schematic structure of lead borate glasses: ●) B; ○) O; ◉) Pb.

It may be concluded that at temperatures below the softening range relaxation losses must always be greater than conduction losses. For the glasses investigated this conclusion is valid at temperatures sufficiently far from the softening range. Data for the glass of the composition (mole %): 70.8B_2O_3, 29.2PbO are presented as an illustration (Table 2).

In this table $\tan \delta_{rel}$ and $\tan \delta_{\sigma}$ represent tangents of the relaxation and conduction loss angles, respectively.

It is clear from the table that at low temperatures relaxation losses are very much greater than conduction losses. However, the conduction losses increase more rapidly than relaxation losses with rise of temperature, and become predominant in the softening range.

Plots of the function

$$\log \tan \delta = f\left(\frac{1}{T}\right) \tag{3}$$

for the glasses studied are given in Fig. 3. The plot for each glass is a broken line consisting of two straight portions; each portion is represented by an equation of the form

$$\tan \delta = Be^{-U/kT} \tag{4}$$

where U is the activation energy of the ion.

TABLE 2

	Temperature, °C						Change
	189	200	225	250	275	300	
$\tan \delta_{rel}$	$1.21 \cdot 10^{-3}$	$1.69 \cdot 10^{-3}$	$2.88 \cdot 10^{-3}$	$4.24 \cdot 10^{-3}$	$4.96 \cdot 10^{-3}$	$2.95 \cdot 10^{-3}$	2.4-fold
$\tan \delta_{\sigma}$	$1.78 \cdot 10^{-5}$	$3.75 \cdot 10^{-5}$	$9.16 \cdot 10^{-5}$	$4.17 \cdot 10^{-4}$	$1.06 \cdot 10^{-3}$	$5.26 \cdot 10^{-3}$	295-fold
$\frac{\tan \delta_{rel}}{\tan \delta_{\sigma}}$	68	45	31.5	11.5	4.7	0.56	120-fold

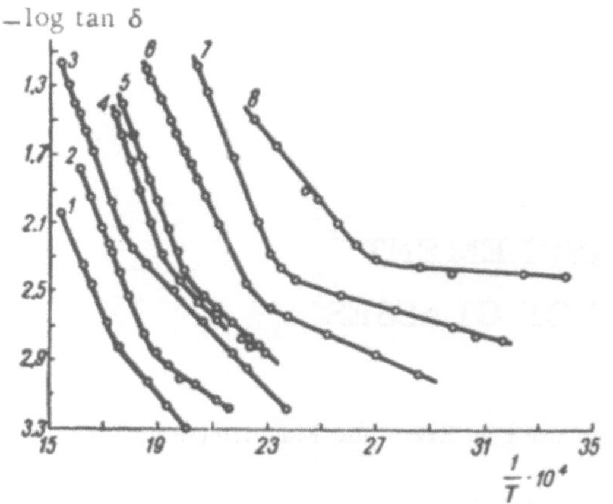

Fig. 3. Analysis of the temperature–loss angle relationship.

Assuming that the losses are due to two types of ions (B and C) simultaneously, with different activation energies U_1 and U_2, we can write the following general expression for the dielectric losses of these glasses:

$$\tan \delta = C_1 e^{-U_1/kT} + C_2 e^{-U_2/kT} \tag{5}$$

where C_2 and C_1 are quantities proportional to the concentrations of ions of types B and C.

For the break point T_{br} we can write the approximate equation

$$C_1 e^{-U_1/kT_{br}} = C_2 e^{-U_2/kT_{br}}$$

whence

$$\frac{C_1}{C_2} = \frac{N_1}{N_2} = e^{(U_1 - U_2)/kT_{br}} \tag{6}$$

Here N_1 and N_2 are the numbers of ions of types C and B per unit volume of the glass.

TECHNIQUE FOR MEASUREMENT OF THE CONDUCTIVITY OF GLASSES

The discussion of the method for conductivity measurement used by Zorin and Mazurin (see p. 104) is summarized below.

V. A. Ioffe. When a potential difference is applied to a glass specimen, current which decreases with time flows through the glass. If we short-circuit the electrodes after some time, we detect a current flowing in the opposite direction. Therefore two electromotive forces act in the specimen. The direction and magnitude of one are determined by the external source of current, and the other arises within the specimen. It is immaterial whether the charge separation giving rise to the induced emf occurs within the specimen or on its surface. In either case the circuit obeys Ohm's law, according to which the current strength in the circuit is equal to the algebraic sum of the emf's divided by the algebraic sum of the resistances.

A. F. Ioffe* showed that the change of current in the circuit is due to an emf of opposite sign which arises in the circuit. If this emf is taken into account, the same value is obtained for the conductivity regardless of the determination conditions.

The residual current depends on the polarization emf, which itself is a function of the field intensity in the specimen. Therefore the conductivity calculated from the residual current must vary with the voltage applied to the specimen, and it cannot be regarded as a physical quantity.

A. P. Zorin and O. V. Mazurin. Our experiments showed that when conductivity is measured in absence of polarization effects (when the current does not decrease with time) the results of the conductivity determinations are not changed when the applied voltage is changed by a factor of 10^3.

It is well known that any local displacement of charges, due to electronic, ionic, or thermionic polarization, in a dielectric placed in an electric field gives rise to an emf of the opposite sign. When the field is removed and the electrodes are short-circuited the return of the charges to their original positions produces a current of the opposite sign in the conductor.

However, it is also known that when a direct voltage is applied to a dielectric it is the difference between the direct emf, determined by the charge on the electrodes, and the reverse emf, determined by polarization effects, which is equal to the voltage of the external source. This difference between the two emf's, measured by means of a voltmeter or electrometer connected to the electrodes, is used for calculating the conductivity.

A. F. Ioffe's techniques must be used only when the influence of the emf due to polarization at the electrodes must be excluded. Our determinations are carried out under conditions when the influence of this emf is negligible.

K. S. Evstrop'ev. In investigations carried out by L. Yu. Kurtts in the State Optical Institute it was shown that polarization effects (although very transient) also occur at high temperatures. Nevertheless, nobody imagines that in high-temperature measurements some sort of correction must be applied for zero time with polarization effects taken into account. Reliable and reproducible results are obtained without this.

It is therefore logical to perform conductivity determinations at low temperatures at time instants when the polarization processes have been completed.

---

*A. F. Ioffe, Physics of Crystals, Gosizdat (1929).

If the problem is to study polarization specially, the techniques of A. F. Ioffe's school should be used.

To avoid misunderstandings, we should not speak of specific conductance during the times when the current passing through the glass varies as the result of polarization processes.

B. I. Sazhin. In organic polymers the current also depends distinctly on time. If we use the loss theory to calculate from the current–time relation the loss–frequency relation in the $10\text{-}10^5$ cps range we obtain a curve which, within the limits of experimental error, is a continuation of the experimental curve for higher frequencies. It is assumed in such calculations that the reverse emf due to polarization is zero.

On the other hand, some materials exhibit a clear dependence of conductivity on field strength under conditions where Poole's law can hardly be expected to apply.

The concept of polarization emf therefore requires clarification.

III

INVESTIGATION OF GLASSES WITH ELECTRONIC CONDUCTION

ELECTRICAL CONDUCTIVITY OF GLASSES IN SYSTEMS As–Ge–Pb–Se

V. A. Khar'yuzov and A. M. Efimov

The glasses studied were based on the system As–Ge–Se, which was the subject of earlier investigations [1, 2]. Information is also available on the conductivities of two-component glasses in this system: arsenic–selenium [3, 4] and germanium–selenium [5, 6].

In the system As–Ge–Pb–Se the region of glass formation becomes narrower with increasing lead content and becomes negligible when it exceeds 5 atomic %. In the present investigation compositions were synthesized mainly with 5 atomic % Pb in the two sections with 15 atomic % As and 60 atomic % Se, and also the corresponding compositions in the ternary system. The glasses were made in sealed quartz tubes by a known method [7]. The conductivities of the glasses were determined in the temperature range of 20-320°C [5].

The conductivity–temperature plots (log σ versus 1/T) for most of the glasses are linear throughout the range investigated (Fig. 1). Only a few glasses in the system As–Ge–Se in the composition range (atomic %): 10-20 As, 25-30 Ge, and 55-60 Se showed considerable deviations in the positions of the experimental log σ versus 1/T plots corresponding to heating of the specimen to 200-240°C and cooling from that temperature (curve 14 in Fig. 1). The observed deviations are due to additional annealing of the glass during conductivity determinations at 200-240°C. This apparently results in irreversible structural changes which affect the conductivity of the glasses. The suggestion that these conductivity changes are associated with crystallization of the glass has not been confirmed.

The results for glasses containing 15 atomic % As are given in Fig. 2. The values of log σ_0 for the glasses are in the range of 3.3-4.5. The variation of log σ with concentration is represented by a curve with a minimum. The minimum for ternary glasses (curve 2) corresponds to approximately 26 atomic % Ge, and for glasses containing lead (curve 1) to 20 atomic % Ge. The minima of these curves correspond to the maxima of the ε_σ curves.

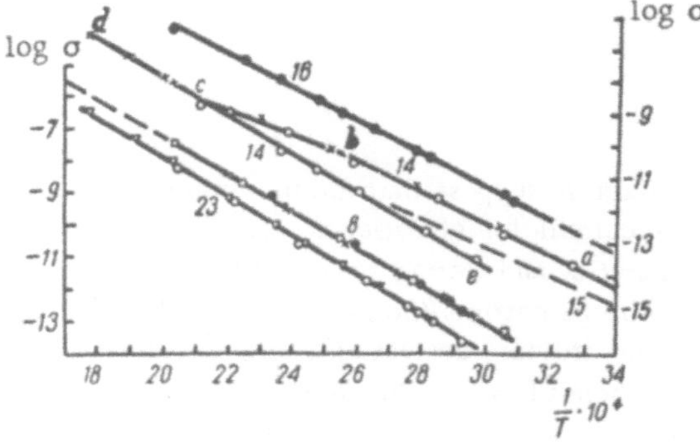

Fig. 1. Conductivity–temperature relations for the glasses (data for specimens from duplicate meltings are distinguished by the symbols at the points). abcd) Heating of a specimen of glass No. 14 of the composition (mole %): 15 As, 30Ge, 55Se; dce) cooling and repeated determinations for a specimen of glass No. 14; 15) glass of the same composition, from data in [1].

Thus, introduction of lead merely leads to displacement of the extremal points on the curves without changing their form.

A minimum of the plot of the total volume concentration of atoms versus the germanium content was found earlier [2] for ternary glasses; it was shown that the minimum corresponds to the composition in which all the arsenic forms $AsGeSe_{5/2}$ structural elements. An analogous minimum, which can probably be attributed to accumulation of the same $AsGeSe_{5/2}$ structural elements, is observed for the glasses in the system As–Ge–Pb–Se under investigation (see Fig. 3). At the same time, comparison of Figs. 2 and 3 shows that the extremal points of the plots of ε_σ, log σ, and total atomic concentration versus the composi-

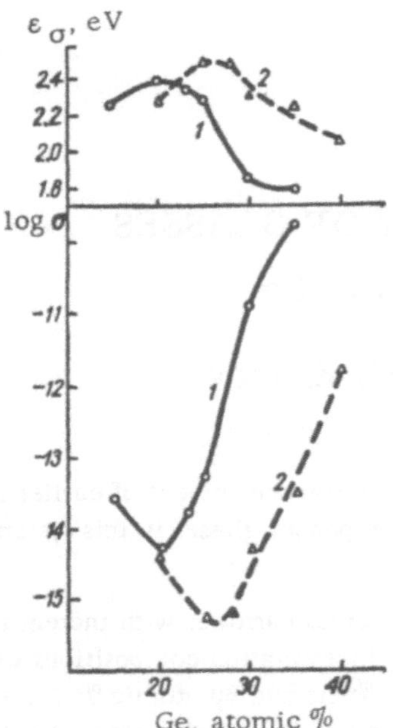

Fig. 2. Variations of conductivity σ (at 50°C) and activation energy ε_σ with glass composition in the section with 15 atomic % As. 1) Glasses of the system As–Ge–Pb–Se with 5 atomic % Pb; 2) glasses of the system As–Ge–Se. The final (steady) values are given for glasses which showed changes of conductivity when heated to 200-240°C. The activation energy was calculated from the formula $\sigma = \sigma_0 \cdot \exp(-\varepsilon_\sigma/2kT)$.

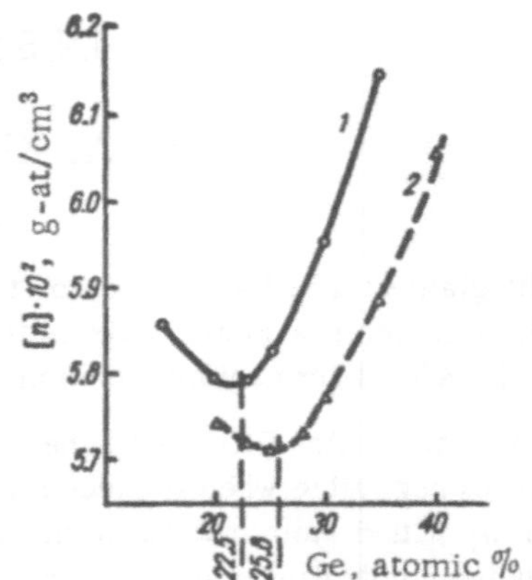

Fig. 3. Total concentration of atoms per cm^3 as a function of composition of glasses containing 15 atomic % As. 1) Glasses of the system As–Ge–Pb–Se; 2) glasses of the system As–Ge–Se, from data in [2].

tion approximately coincide for each of the systems. We may therefore suppose that the maximum of ε_σ and the corresponding minimum of log σ may also be attributed to accumulation of $AsGeSe_{5/2}$ structural elements. These structural elements form chains with alternating As–Se, As–Ge, and Ge–Se bonds, differing considerably in their ionization energies ε_i. It may be supposed that the charge carriers (holes) traveling along such a chain must surmount a potential barrier which includes not only the ordinary activation energy for displacement ε_a, but also the difference between the ionization energies of the bonds forming the chain $\Delta\varepsilon_i = \varepsilon_{i\,Ge-Se} - \varepsilon_{i\,As-Ge}$. The appearance of this term in the magnitude of the activation barrier accounts, in our opinion, for the maximum of ε_σ at the maximum content of $AsGeSe_{5/2}$ structural elements.

The dependence of conductivity on composition in the system As–Ge–Se is analogous to that found in [1], but the absolute values of ε_σ and log σ_0 for glasses of similar composition differ appreciably (by 0.2-0.5 eV for ε_σ and by 2-3 units for log σ_0).

For all the glasses investigated by us, log β (where β is the steric factor [8]) is between −0.2 and +1.0; this is evidence of almost complete agreement between the experimental and theoretical values of log σ_0.

Literature Cited

1. R. L. Myuller, L. A. Baidakov, and Z. U. Borisova, Vestn. Leningr. Gos. Univ.,Ser. Fiz. i Khim. (10) part 2:94-102 (1962).
2. L. G. Aio and V. F. Kokorina, Optiko-Mekh. Prom. (2):36 (1963).
3. L. A. Baidakov, Z. U. Borisova, and R. L. Myuller, Zh. Prikl. Khim.,34:2446 (1961).
4. K. S. Evstrop'ev, O. V. Mazurin, and V. A. Khar'yuzov, Tr. Leningr. Tekhnol. Inst. im Lensoveta (152):16 (1961).
5. V. A. Khar'yuzov and K. S. Evstrop'ev, Optiko-Mekh. Prom. (10):17-21 (1961).
6. Z. U. Borisova, R. L. Myuller, and Chin Cheng-ts'ai, Zh. Prikl. Khim.,35(4):774 (1962).
7. L. G. Aio and V. F. Kokorina, Optiko-Mekh. Prom. (4):39 (1961).
8. R. L. Myuller, Zh. Prikl. Khim.,35(3):541-550 (1962); Vestn. Leningr. Gos. Univ.,Ser. Fiz. i Khim.(22) part 4:86-96 (1961).

ELECTRICAL CONDUCTIVITY OF CHALCOGENIDE GLASSES

Brief abstracts of earlier publications on the electrical conductivity of chalcogenide glasses are given below.

L. A. Baidakov, Z. U. Borisova, and R. L. Myuller. Conductivity of glassy AsS_x dielectrics with electronic conduction.

The conductivity of the glass compositions studied, with $1.25 \leq x \leq 18$, at 20-100°C is in the range of 10^{-10}-10^{-18} $ohm^{-1} \cdot cm^{-1}$. The experimental data are satisfactorily explained by the valence nature of the electronic conduction.

[Vestn. Leningr. Gos. Univ. (22):77-89 (1962)].

L. A. Baidakov. Conductivity of the glassy semiconducting system As—Ge—Se.

Experimental data on the energy and modulus of conductivity of this system indicate that through conduction is limited by a network skeleton of As—Se and Ge—Se bonds.

[Vestn. Leningr. Gos. Univ. (22):105-113 (1962)].

Z. U. Borisova, L. A. Baidakov, V. V. Ipat'eva, and G. A. Chernova. Conductivity of the glassy system As—S—Se with a mixed semiconductor—dielectric function.

Experimental data on conductivity indicate a smooth change of properties as selenium is replaced by sulfur, in agreement with the valence hypothesis of electrical conduction in glasses.

[Vestn. Leningr. Gos. Univ. (22):90-95 (1962)].

Z. U. Borisova and A. I. Bobrov. Conductivity of glassy $As \cdot Se_xGa_y$.

Introduction of gallium into $As-Se_x$ glasses raises the conductivity by two to three orders of magnitude; this is attributed to stabilization of through transport of current carriers.

[Vestn. Leningr. Gos. Univ. (22):159-164 (1962)].

R. L. Myuller and E. V. Shkol'nikov. Investigation of crystallization of $AsSe_xGe_y$ glasses by conductivity measurements.

High-temperature annealing of $AsGe_{2.33}Se_{3.0}$ and $AsGe_{4.0}Se_{5.0}$ glasses resulted in volume crystallization with separation of a GeSe phase. The glasses were chilled and conductivity was determined at low temperatures. It was found that the energy of conduction fell during crystallization from 1.5-1.7 eV to 0.3-0.5 eV, with a relatively small change in the conductivity modulus. The crystallization rate constants were estimated and the activation energy of the process was calculated.

[Vestn. Leningr. Gos. Univ. (22):119-133 (1962)].

V. P. Pozdnev. Investigation of the conductivity of semiconductor glasses in the system As_2Se_3—As_2Te_3 in the liquid state by a noncontact method.

Conductivity measurements on glasses of this system by a noncontact method at 500-800°C give values higher by 5-6 orders of magnitude than those obtained by the usual contact method. The author attributes this effect to the presence of a considerable amount of crystallites with high conductivity in the melt.

[Fiz. Tverd. Tela.,4:946 (1962)].

ELECTRICAL CONDUCTIVITY OF THALLIUM SULFOSELENIDE–ARSENIC GLASSES IN RELATION TO THEIR STRUCTUROCHEMICAL CHARACTERISTICS

T. P. Markova

Despite the similarity of their valence electron shells, the glasses formed when sulfur and selenium are fused with arsenic and thallium differ greatly in their electrical properties [1, 2]. It has been found that the conductivity of glassy alloys in the system As–Se–Tl at 20°C is in the range of 10^{-6}-10^{-12} $ohm^{-1} \cdot cm^{-1}$, and the energy ε_{σ} = 1.0-1.4 eV [3]. The conductivity of glasses in the system As–S–Tl is less by 6-10 orders of magnitude, and the energy ε_{σ} = 1.5-1.8 eV [4].

It was important to elucidate in detail how the conductivity of thallium–arsenic–selenium glasses alters when selenium is gradually replaced by sulfur.

The specimens were prepared by vacuum fusion of redistilled arsenic and selenium, sulfur of A-2 grade, and thallium of V-2 grade. The methods used for measuring conductivity, microhardness, and specific gravity were the same as before [3, 4]. The compositions chosen for the investigation had the general formula $AsS_xSe_{z-x}Tl_y$ with $1.9 \leq z \leq 2.3$ and $0.3 \leq y \leq 1.2$, with moderate excess of sulfur and selenium $[z-(1.5+0.5y) \leq 0.6$,

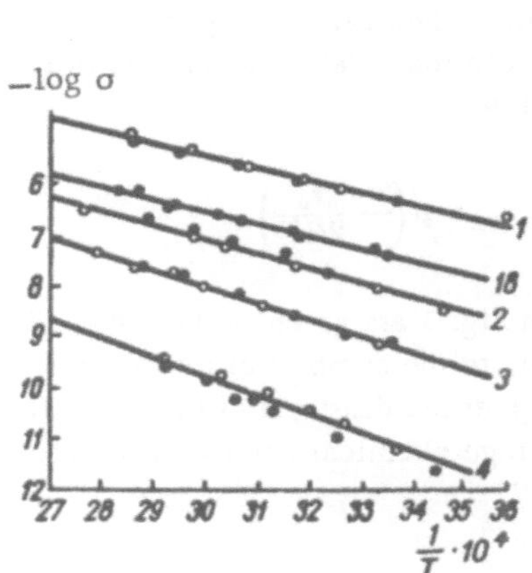

Fig. 1. Conductivity–temperature relations of glassy $AsS_xSe_{z-x}Tl_y$ alloys (the numerals correspond to the glass numbers).

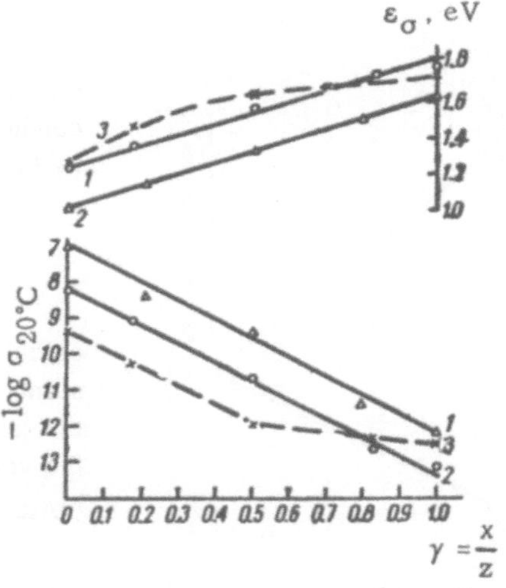

Fig. 2. Changes of conductivity and energy of conduction when selenium is replaced by sulfur in $AsS_xSe_{z-x}Tl_y$ glasses. 1) z = 1.9; y = 0.8; 2) z = 2.3; y = 0.5; 3) z = 2.3; y = 0.3.

Glass No.	Average chemical composition $AsS_xSe_{z-x}Tl_y$			Mol. wt. M	Density d, g/cm³	Content [M] 10², moles/cm³	$\gamma=\frac{x}{z}$	−log σ at 20°C	Energy ε_σ, eV	log σ_{0e}	log $\frac{\sigma_{0e}}{[v]}$	log β
	z	x	y									
1	1.9	0	0.8	388	6.30±0.01	1.6	0.00	6.5	1.00	2.0	3.3	—1.3
2	1.9	0.4	0.8	370	5.92±0.01	1.6	0.21	8.4	1.13	1.4	2.7	—1.9
3	1.9	0.95	0.8	344	4.67±0.03	1.4	0.50	9.4	1.32	1.9	3.2	—1.4
4	1.9	1.5	0.8	318	4.98±0.01	1.6	0.72	11.4	1.51	1.5	2.8	—1.8
5	1.9	1.9	0.8	299	5.18±0.01	1.7	1.00	12.2	1.64	1.6	2.9	—1.7
6	2.26	0	1.0	458	6.32±0.01	1.4	0.00	6.5	1.12	3.0	4.3	—0.3
7	2.26	0.13	1.0	405	5.77±0.02	1.6	0.50	8.3	1.20	2.0	3.3	—1.3
8	2.3	0	0.3	318	5.24±0.01	1.7	0.00	9.4	1.26	1.4	2.6	—2.0
9	2.3	0.4	0.3	299	5.0 ±0.1	1.7	0.17	10.3	1.46	2.1	3.3	—1.3
10	2.3	1.15	0.3	264	4.55±0.01	1.7	0.50	12.0	1.64	2.0	3.2	—1.4
11	2.3	1.9	0.3	229	4.6 ±0.1	2.1	0.83	12.4	1.67	1.8	3.0	—1.6
12	2.3	2.3	0.3	210	3.77±0.01	1.8	1.00	12.5	1.76	2.3	3.5	—1.1
13	2.3	0	0.5	359	5.61±0.01	1.6	0.00	8.3	1.23	2.3	3.5	—1.1
14	2.3	0.4	0.5	340	5.35±0.03	1.7	0.18	9.1	1.36	2.5	3.7	—0.9
15	2.3	1.15	0.5	305	4.96±0.01	1.6	0.50	10.7	1.57	2.7	3.9	—0.6
16	2.3	1.9	0.5	270	4.49±0.01	1.7	0.83	12.6	1.75	2.4	3.6	—1.0
17	2.3	2.3	0.5	251	4.25±0.01	1.7	1.00	13.1	1.79	1.3	2.5	—2.1
18	2.0	1.0	1.2	431	5.6 ±0.1	1.3	0.50	7.6	1.00	1.0	2.4	—2.2

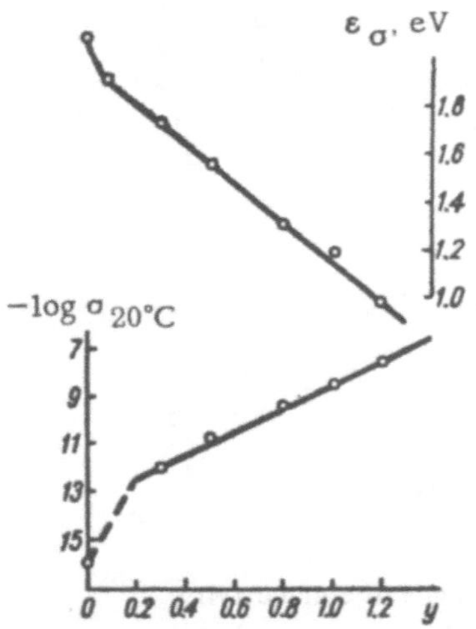

Fig. 3. Variations of conductivity and energy of conduction of $AsS_xSe_{z-x}Tl_y$ glasses with the thallium content, with [S]/[Se] = z/(z − x) = 1.

The experimental conductivity data are plotted in Fig. 1. Specimens from different meltings show satisfactory reproducibility of the conductivity–temperature relation. Values of ε_σ and of the statistical factor log σ_{0e} in the equation

$$\sigma=\sigma_{0e}\exp\left(-\frac{\varepsilon_\sigma}{2RT}\right)$$

calculated from the data in Fig. 1 are given in the table. The curve numbers in Fig. 1 correspond to the numbers of the compositions in the table. The table also gives the density d of the glasses (g/cm³) and the volume contents of average chemical units with molecular weight $M_{AsS_xSe_{z-x}Tl_y}$:

$$[M]=\frac{d}{M_{AsS_xSe_{z-x}Tl_y}}$$

The microhardness of $AsS_xSe_{z-x}Tl_y$ glasses varies little; the average value is (120 ± 10) kg/mm². Figure 2 represents changes in the logarithm of the conductivity (at 20°C) and in the energy of conduction during progressive replacement of selenium by sulfur in glasses with $1.9 \le z \le 2.3$; $0.3 \le y \le 0.8$. It is clear from Fig. 2 that as selenium is replaced by sulfur the energy of conduction increases from 1.0 to 1.8 eV while the conductivity decreases from 10^{-7} to 10^{-13} $ohm^{-1} \cdot cm^{-1}$; in the first approximation the changes are directly proportional to the degree of replacement of selenium by sulfur: $-\log \sigma_{20°C} = a + 5.15\gamma$; $\varepsilon_{\sigma} = b + 0.6\gamma$ where $\gamma = x/z$ is the degree of replacement of selenium by sulfur. When z = 1.9, y = 0.8, a = 7.0, b = 1.02; when z = 2.3, y = 0.5, a = 9.4, b = 1.23. This simple relation between conductivity and composition becomes more complicated for glasses with z = 2.3, y = 0.3, probably as the result of phase separation when selenium is replaced by sulfur.

In Fig. 3 log conductivity (at 20°C) and the energy of conduction are plotted against the thallium content with 50% replacement of selenium by sulfur, for $1.9 \leq z \leq 2.3$. The abscissa gives the coefficient y in the formula of the glass. As before, variations of ε_{σ} and $\log \sigma_{20°C}$ with the atomic content of thallium are almost linear: $-\log \sigma_{20°C} = 13.4 - 4.9y$; $\varepsilon_{\sigma} = 1.98 - 0.83$ y.

For all the glasses studied, $\log \beta \approx -1.3$. This indicates the absence of any appreciable hindrance to through transfer of charge. This probably accounts for the simple relation found between the conductivity and the composition of these glasses.

The author is deeply grateful to Professor R. L. Myuller for his guidance and to S. V. Nemilov for valuable comments in discussion of the experimental results.

Literature Cited

1. L. A. Baidakov, Z. U. Borisova, and R. L. Myuller, Zh. Prikl. Khim., 34:2446 (1961).
2. R. L Myuller, L. A. Baidakov, and Z. U. Borisova, Vestn. Leningr. Gos. Univ., (22):77 (1962).
3. R. L. Myuller and T. P. Markova, Vestn. Leningr. Gos. Univ. (4):75 (1962).
4. T. P. Markova, Vestn. Leningr. Gos. Univ. (22):96 (1962).

SEMICONDUCTING SILICATE GLASSES

A. Ya. Kuznetsov and V. A. Tsekhomskii

Investigations of semiconducting silicate glasses based on iron oxides have been fairly numerous [1, 2, 3]. The occurrence of electronic conduction in these glasses is due to the equilibrium

$$Fe^{2+} \rightleftarrows Fe^{3+} + e^-$$

which arises when the glass is melted at 1400-1500°C. Melting under strongly oxidizing or strongly reducing conditions leads to decreased conductivity as the result of a sharp shift of the equilibrium in one direction or the other. This is clear from Fig. 1, which shows temperature–resistance relations for four glasses containing equal amounts of iron oxides (2 mole %) but made under different oxidation–reduction conditions. Glasses No. 2 and No. 3 were made with 2 and 5% of carbon in the batch. Barium oxide was introduced into glass batch No. 4 in the form of nitrate (i.e., oxidizing conditions were created). Figure 1 shows that these glasses have lower conductivities than glass No. 1, made under the normal conditions.

With the influence of oxidation or reduction conditions taken into account, all the glasses investigated were melted at 1300-1550°C under the usual conditions in an oil-fired furnace in 3-liter quartz pots or in a laboratory silit furnace in 200-ml quartz or corundum crucibles. The compositions and electrical characteristics of these glasses are given in the table.

Introduction of iron oxides into barium and lead silicate glasses raises the conductivity sharply. Lead silicate glasses have higher conductivities in all cases than barium silicate glasses of equal Fe_2O_3 content. This difference can be explained as follows. For electronic transport the glass must contain trivalent iron, but electron transfer appears possible only through Fe^{3+} ions in sixfold coordination, because trivalent iron in fourfold coordination has the following structural form:

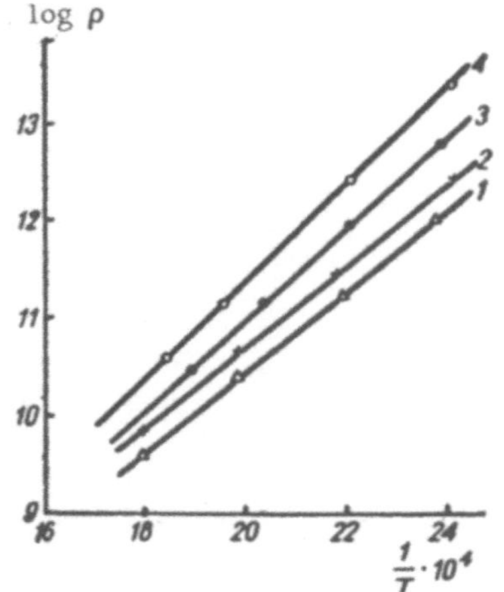

Fig. 1. Resistance–temperature relations of glasses made under various conditions (see text).

```
   O   Me+
   |
O—Fe⁻—O
   |
   O
```

In this tetrahedron the iron ion is negatively charged and approach of an electron is improbable. The glass must contain weakly bonded oxygen atoms to allow iron to acquire fourfold coordination. The content of such atoms increases with decrease of the field of the modifying ion. Since the radius of the lead ion (1.32 A) is less than that of the barium ion (1.43 A), lead glasses have higher conductivity than barium glasses. The decrease of conductivity with increasing content of the modifying oxide in the glass can be explained similarly (Fig. 2). Figure 2 shows that, with the same iron oxide content (10 mole %), the conductivity decreases tenfold when the PbO content is doubled (from 30 to 60 mole %).

Glass No.	SiO_2	Al_2O_3	PbO	BaO	CaO	Fe_2O_3	$\log \rho_{150}$	$\log \rho_{300}$	E, eV
1	47.5	—	47.5	—	—	5	9.7	7.4	0.67
2*	45	—	45	—	—	10	7.6	5.5	0.68
2	45	—	45	—	—	10	8.1	6.0	0.68
3	60	—	30	—	—	10	7.6	5.6	0.64
4	50	—	40	—	—	10	7.9	5.8	0.67
5*	30	—	60	—	—	10	8.7	6.6	0.67
6	57.5	—	—	37.5	—	5	10.7	8.1	0.83
7	55	—	—	35	—	10	8.9	6.6	0.74
8	69	—	—	21	—	10	8.4	6.1	0.73
9	45	—	—	45	—	10	9.2	6.8	0.77
10	63	9	—	—	26	2	14.3	11.2	0.99
11	60	9	—	—	26	5	11.0	8.4	0.83
12	55	9	—	—	26	10	9.7	7.3	0.77
13	50	9	—	—	26	15	7.8	5.6	0.72
14	65	3	—	—	23	9	8.4	6.0	0.76
15	62	6	—	—	23	9	8.7	6.4	0.73
16	59	9	—	—	23	9	9.7	7.2	0.80
17*	42	3	45	—	—	10	7.5	5.6	0.63
18*	40	5	45	—	—	10	7.7	5.7	0.64
19*	35	10	45	—	—	10	8.6	6.4	0.70
20	59	5	—	27	—	9	9.0	6.8	0.72
21	57	7	—	27	—	9	9.2	7.0	0.72
22	54	10	—	27	—	9	9.0	6.8	0.72

Note. Glasses made in the silit furnace are marked with an asterisk.

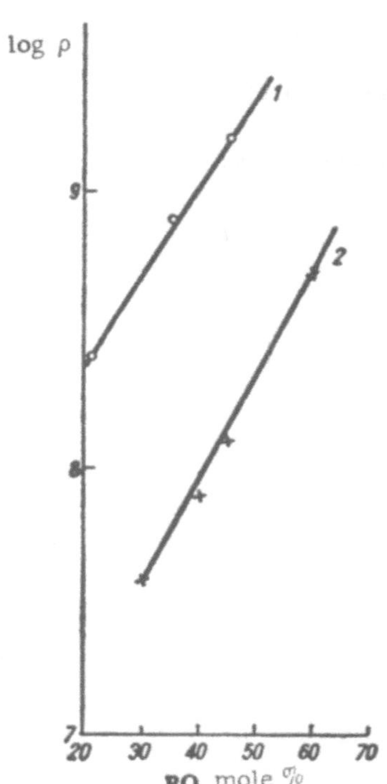

Fig. 2. Effects of modifier-ion content on the resistivity of barium (1) and lead (2) glasses at 150°C.

Figure 3 shows variations of the resistivity of calcium aluminosilicate glasses with increasing iron oxide content. The curve for lead glasses [5] is shown for comparison. According to the above reasoning, calcium glasses should have somewhat higher conductivities than lead glasses. However, Fig. 3 shows that the lead glasses have higher conductivities in all cases. This discrepancy can be attributed to the presence of alumina in the glass. It is known that aluminum oxide replaces ferric oxide isomorphously in the crystal lattice, and this is accompanied by "dilution" of iron ions by aluminum ions not involved in electronic charge transfer. Another consequence of "dilution" by aluminum ions is that the equilibrium between Fe^{3+} ions in six- and fourfold coordination shifts in favor of four-coordinated iron [5]. As has already been shown, electron transport appears possible only through iron ions in sixfold coordination.

It must be pointed out that the influence of aluminum oxide on conductivity of iron-containing glasses depends greatly on the modifying ion (Fig. 4). In the case of the large barium ion introduction of aluminum oxide has very little effect on conductivity. In lead glasses up to about 3% of aluminum oxide has no effect on conductivity, but starting with 3% the resistivity rises fairly sharply.

Although glasses consisting of calcium oxide and silica could not be obtained, because of the high melting temperatures and the strong tendency to crystallization, it is quite clear from Fig. 4 that increase of the aluminum oxide content above 3% raises their resistance sharply.

Alumina "dilutes" iron oxides in glasses, decreasing the conductivity, if the modifying ion is comparable in size to the Fe^{2+} ion. If the radius of the modifying ion is much greater than that of the Fe^{2+} ion, the aluminum ion interacts with the more loosely bound oxygen at the modifying ion and has no effect on conductivity. Similar results were obtained earlier for alkali aluminosilicate glasses [4].

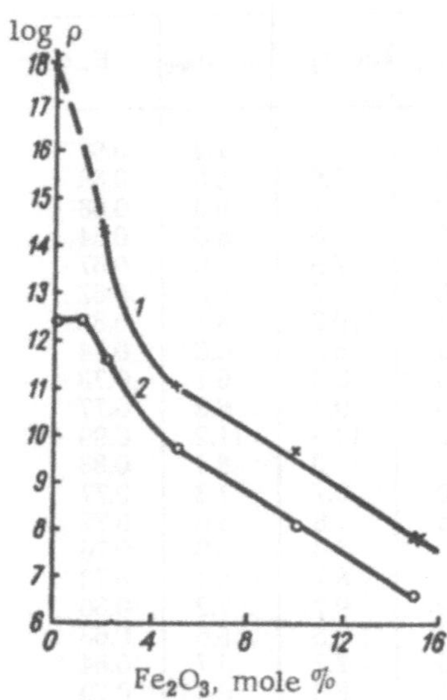

Fig. 3. Variations of the resistivity of calcium aluminosilicate (1) and lead silicate (2) glasses with the Fe_2O_3 content at 150°C.

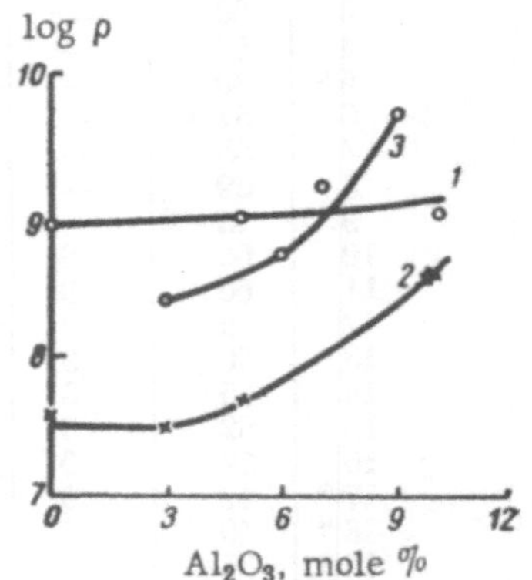

Fig. 4. Variations of the resistivity of barium (1), lead (2), and calcium (3) glasses with the Al_2O_3 content at 150°C.

Literature Cited

1. O. V. Mazurin, G. A. Pavlova, E. Ya. Lev, and E. K. Leko, Zh. Tekhn. Fiz., 27(12):2702 (1957).
2. A. Ya. Kuznetsov and V. A. Tsekhomskii, Optiko-Mekh. Prom. (7):27 (1962).
3. L. A. Grechanik, E. A. Fainberg, and I. N. Zertsalova, Fiz. Tverd. Tela, 4(2):454 (1962).
4. K. K. Evstrop'ev and V. A. Tsekhomskii, Fiz. Tverd. Tela, 4(12):3390 (1962).
5. G. O. Karapetyan, V. A. Tsekhomskii, and D. M. Yudin, Fiz. Tverd. Tela, 5(2):627 (1963).

STUDY OF SEMICONDUCTING IRON-CONTAINING GLASSES BY ELECTRON PARAMAGNETIC RESONANCE

G. O. Karapetyan, V. A. Tsekhomskii, and D. M. Yudin

The electron paramagnetic resonance (EPR) method gives information on the state of a central paramagnetic ion and its first coordination sphere. The technique for recording EPR spectra was described by us earlier [1]. There is no long-range order in glasses, and their conductivity is determined by the properties of the first coordination spheres of elements which cause conduction. In iron-containing glasses iron is such an element. The electron configuration of the iron atom is $3d^6 4s^2$, and its ions have configurations $3d^6$ (Fe^{2+}) and $3d^5$ (Fe^{3+}). Bivalent iron does not give an EPR signal at temperatures above 77 K. The trivalent iron ion gives an EPR spectrum in the form of several resonance lines, which may be present either separately or together in glass. The field of the first coordination sphere and the corresponding g factors were calculated by Castner [2], who investigated sodium calcium silicate glasses containing iron (0.1% by weight).

We investigated various iron-containing alkali-free lead silicate and barium silicate glasses; the effects of boron and aluminum and the influence of oxidation and reduction conditions during the melting were studied. The EPR investigations were accompanied by studies of optical absorption spectra, and spectral changes were correlated with conductivity variations. In some cases resonance absorption of EPR spectra begins in the zero field region and continues beyond 5000 Oe. The EPR spectra of lead silicate glasses were published earlier [3]. Lines with $g = 6$ and $g = 4.3$ disappear from the EPR spectra with increase of Fe^{3+} concentration, and the line with $g = 2$ remains and increases in intensity. It is due to the presence of Fe^{3+} in sixfold coordination; this is confirmed by the following experiment. An Fe^{3+} ion in sixfold coordination and acting as a modifier should have weaker interaction with the glass lattice than an Fe^{3+} ion in tetrahedral coordination. It is known that interaction of paramagnetic ions with the lattice, known as spin–lattice interaction, leads to broadening of the resonance lines in the EPR spectrum. If we record the EPR spectrum at room temperature, cool the glass, and again record the EPR spectrum, the resonance lines of the network-forming ions should become narrower and their resolution should increase. An experiment with lead silicate glass showed (see the figure) that the line in the region of $g = 2$ shows no substantial change when the glass is cooled to 77 K, and only the resolution of the line with $g = 4.3$ increases. These results lead to the tentative conclusion that as the iron concentration increases its role as a modifier also increases. In lead silicate glasses the equilibrium between iron ions in fourfold and sixfold coordination is shifted toward sixfold coordination in comparison with barium silicate glasses; this follows from the EPR spectra [3]. Accordingly, lead silicate glasses have higher conductivities. Increase of the amount of reducing agent (carbon) in the batch reduces the total intensity of the EPR spectrum owing to the transition $Fe^{3+} \rightarrow Fe^{2+}$. The line with $g = 2.0$ decreases relatively more rapidly than the line with $g = 4.3$. At the same time, under the intensified reducing conditions in the melting, the conductivity is decreased by a factor of 2.5, although the Fe^{3+} ion concentration increases. This leads to the conclusion that electronic conduction in iron-containing glasses requires the presence of trivalent iron in sixfold coordination. The effects of aluminum oxide in lead silicate glasses, and of replacement of SiO_2 by B_2O_3, are interesting. Both the oxides, Al_2O_3 and B_2O_3, increase the resistance and the activation energy and raise light absorption sharply. Analysis of the EPR spectra shows that addition of aluminum and replacement of SiO_2 by B_2O_3 increase the number of Fe^{3+} ions in tetrahedral coordination; iron in fourfold coordination has greater absorption than in sixfold coordination. It is quite natural that conductivity is associated with Fe^{3+} ions in sixfold and not fourfold coordination. The $[Fe^{3+}O_4]$ tetrahedron has at least one negative charge and approach of electrons to it is difficult.

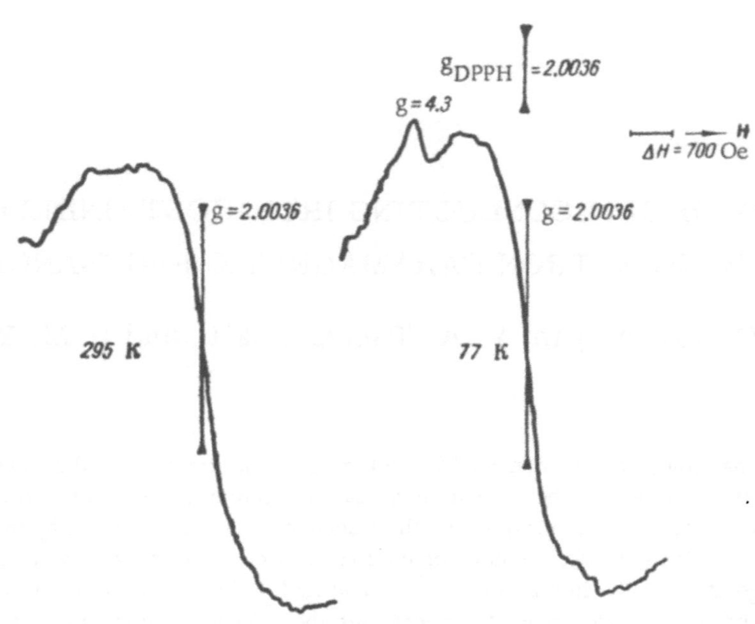

EPR spectra of lead silicate glass of the composition (mole %): $10FeO_3 \cdot 30PbO \cdot 60SiO_2$, recorded at 295 K and 77 K.

Literature Cited

1. G. O. Karapetyan and D. M. Yudin, Fiz. Tverd. Tela, 3(9):2827 (1961).
2. T. Castner, G. S. Newell, W. C. Holton, and C. P. Slichter, J. Chem. Phys., 32(6):668 (1960).
3. G. O. Karapetyan, V. A. Tsekhomskii, and D. M. Yudin, Fiz. Tverd. Tela, 5(2):627 (1963).

INFLUENCE OF CHEMICAL COMPOSITION ON ELECTRICAL RESISTANCE OF IRON-CONTAINING GLASSES

I. N. Zertsalova

The methods of making the glasses and measuring their electrical resistance are described in the paper by L. A. Grechanik et al.*

The experimental results are given in the tables and the figure. The activation energy of conduction was calculated from the formula

$$\rho = \rho_0 e^{\frac{E}{2kT}}$$

Examination of the experimental material reveals the following features.

The variations of resistance of silicate and borate glasses with the Fe_2O_3 content are similar in character: the resistance falls sharply on introduction of iron ions.

TABLE 1. Compositions and Resistivities of the Silicate Glasses Studied

Glass No.	Chemical comp., mole %†				log ρ at 200°C	E, eV
	CaO	RO	RO_2	Fe_2O_3		
1	33.3	—	—	11.2	7.33	1.31
2	34.3	—	—	14.3	5.85	1.32
3	29.5	—	—	17.0	5.40	1.18
		SrO	**MnO_2**			
4	7.0	38.0	12.5	2.0	12.32	2.31
5	7.0	38.0	8.5	7.0	9.01	1.57
6	7.0	38.0	6.0	8.5	8.56	1.60
7	5.0	28.0	8.0	8.5	8.24	1.50
		PbO				
14	—	22.0	—	8.0	6.87	1.39
16	—	62.0	—	8.0	7.55	1.50
		BaO	**TiO_2**			
17	—	26.0	44.0	—	12.50	2.43
18	—	25.9	43.8	1.0	12.04	1.91
19	—	25.7	43.3	2.0	10.70	1.72
20	—	25.6	43.4	2.5	9.44	1.52
21	—	25.4	42.9	3.0	8.27	1.24
22	—	24.8	42.1	5.0	6.88	1.09
23	—	30.0	25.0	3.0	10.68	1.46
24	—	30.0	25.0	4.0	10.13	1.42
25	—	30.0	25.0	5.0	9.52	1.38
26	—	25.0	20.0	3.0	10.10	1.55

† SiO_2 content found from difference between total and 100%.

*L. A. Grechanik, E. A. Fainberg, and I. N. Zertsalova, Zh. Prikl. Khim., 36(1):91 (1963).

TABLE 2. Compositions and Resistivities of the Borate Glasses Studied

Glass No.	Chemical comp., mole % *			log ρ at 200°C	E, eV
	RO	BaO	Fe_2O_3		
	CdO				
31	35.0	5.0	3.0	12.38	1.98
32	35.0	5.0	5.0	10.84	1.85
33	35.0	5.0	7.0	9.78	1.67
34	35.0	5.0	9.0	8.76	1.62
39	50.0	10.0	7.0	9.16	1.60
40	40.0	10.0	7.0	9.53	1.75
41	30.0	10.0	7.0	10.10	1.86
42	20.0	10.0	7.0	10.40	1.75
45	30.0	40.0	7.0	9.16	1.88
	CaO				
46	37.5	3.2	3.0	11.88	2.00
47	37.5	3.2	5.0	10.56	3.77
48	37.5	3.2	7.0	9.28	1.60
49	37.5	3.2	9.0	8.51	1.56
	ZnO				
50	33.3	—	8.3	8.28	1.30
51	33.3	—	12.5	6.82	1.21

*B_2O_3 content found by difference.

TABLE 3. Compositions and Resistivities of the Phosphate Glasses Studied

Glass No.	Chemical comp., mole %†		log ρ at 200°C	E, eV	Glass No.	Chemical comp., mole %†		log ρ at 200°C	E, eV
	PbO	Fe_2O_3				PbO	Fe_2O_3		
52	30.0	—	4.80	1.37	64	49.0	1.0	8.55	1.89
53	27.0	3.0	6.45	1.69	65	47.0	3.0	8.33	1.83
54	25.0	5.0	7.33	1.97	66	45.0	5.0	7.69	1.65
55	23.0	7.0	7.47	1.97	67	43.0	7.0	7.57	1.66
56	21.0	9.0	8.00	1.73	68	41.0	9.0	7.42	1.50
57	15.0	15.0	7.56	1.36	69	35.0	15.0	6.65	1.31
58	40.0	—	6.50	1.58	70	60.0	—	8.40	1.79
59	39.0	1.0	8.15	1.86	71	59.0	1.0	7.90	1.79
60	35.0	5.0	8.37	1.81	72	57.0	3.0	7.88	1.74
61	31.0	9.0	7.71	1.67	73	55.0	5.0	7.80	1.52
62	25.0	15.0	7.10	1.36	74	53.0	7.0	7.60	1.43
63	50.0	—	7.68	1.84	75	45.0	15.0	6.83	1.37

†P_2O_5 content found by difference.

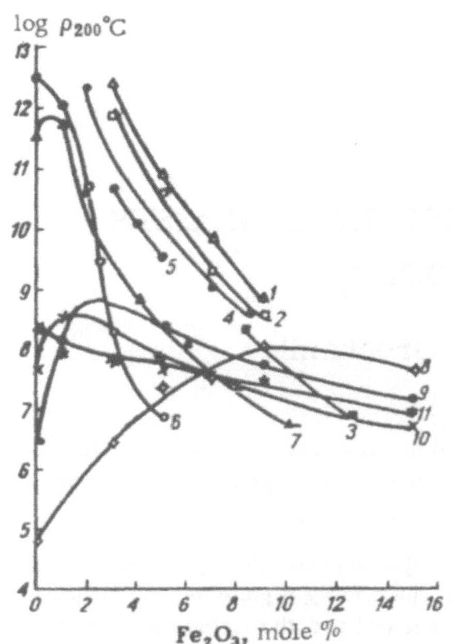

Effect of Fe_2O_3 on the electrical resistance of silicate, borate, and phosphate glasses. 1) Glasses Nos. 31-34; 2) glasses Nos. 46-49; 3) glasses Nos. 50, 51; 4) glasses Nos. 4-6; 5) glasses Nos. 23-25; 6) glasses Nos. 17-22; 7) data from an earlier paper;* 8) glasses Nos. 52-57; 9) glasses Nos. 58-62; 10) glasses Nos. 63-69; 11) glasses Nos. 70-75.

The greatest increases of conductivity on introduction of Fe_2O_3 occur in the systems $SiO_2-PbO-Fe_2O_3$ and $SiO_2-TiO_2-BaO-Fe_2O_3$ at high TiO_2 contents. Increase of the PbO content from 22 to 62% in glasses of the system $SiO_2-PbO-Fe_2O_3$ alters the resistance by less than an order of magnitude, whereas increase of the TiO_2 content from 20 to 45% in glasses of the system $SiO_2-TiO_2-BaO-Fe_2O_3$ lowers the resistance by nearly two orders of magnitude. The effects of other oxides (CaO, SrO, CdO, BaO, MnO_2) on the conductivity of iron-containing silicate and borate glasses are slight: variations of their contents at constant Fe_2O_3 concentration do not produce any significant changes in the resistance.

In the system $P_2O_5-PbO-Fe_2O_3$ the variations of resistance and activation energy with the Fe_2O_3 content alter regularly with increasing content of P_2O_5 in the glass (curves 8-11).

*L. A. Grechanik, E. A. Fainberg, and I. N. Zertsalova, Fiz. Tverd. Tela,4(2):454 (1962).

ELECTRICAL PROPERTIES OF HIGH-LEAD GLASSES REDUCED IN HYDROGEN

E. A. Fainberg and L. A. Grechanik

When high-lead glasses are subjected to heat treatment in hydrogen, a layer of reaction products having a number of peculiar properties is formed on their surface [1-4]. In this paper we give a brief account of the principal results of an investigation of the electrical properties of reduced glasses.

1. The surface electrical resistance of glasses reduced in hydrogen depends greatly on their chemical composition. Investigations of the system $SiO_2-PbO-RO_{0.5}-RO_{1.5}$, where $RO_{0.5}$ represents alkali-metal oxides and $RO_{1.5}$ is $BiO_{1.5}$, $SbO_{1.5}$, $AsO_{1.5}$, or combinations of these oxides, showed that the surface resistance of binary lead silicate glass is not changed as the result of hydrogen treatment at various temperatures. The resistance of the specimens at room temperature is above 10^{12} ohms, although a surface layer of reaction products ranging in color from gray to black is formed. Glasses with added oxides of bismuth, antimony, or arsenic have specific resistances from 10^5 to 10^{10} ohms or higher after reduction.

Alkali-metal ions raise the surface resistance of the glasses; it has been found that the more an alkali-metal ion lowers the volume resistance, the more does it raise the resistance of the reduced layer, and vice versa (Fig. 1). This effect cannot be explained by Blodgett's view of alkali-metal ions as traps which localize free electrons [2], because in that case the effect of alkalies should diminish from lithium to cesium in accordance with weakening of the cation field strength. It appears to be associated in some way with the mobility of the alkali-metal ions in the electric field.

2. Freshly prepared specimens of reduced glasses do not have stable electrical characteristics: their surface resistance rises irreversibly when they are stored under room conditions, especially at elevated temperatures and humidities. It was found that this "aging" of the reduced layers is due not only to chemical interaction with the atmosphere but also to structural changes, which occur even at relatively low temperatures.

Glass compositions and conditions for heat treatment in hydrogen and air have been found which ensure very stable electrical characteristics in the products: the change of resistance does not exceed 4-5% during storage in air for 4 years (Fig. 2), or 2% when the specimens are heated in air or vacuum to 200°C for 100 hr.

3. The temperature coefficient of resistance of hydrogen-reduced layers is negative, and ranges from 0.3 to 1.0% per deg C in accordance with the chemical composition of the original glass and the conditions of heat treatment in hydrogen. The temperature coefficient is almost independent of the reduction time (at constant temperature) and of subsequent heating in air. The activation energy of conduction varies with the temperature.

4. Hydrogen-reduced glass layers have coefficients of secondary electron emission with maximum values of 3.5-5 at primary electron energies of 300-400 V.

5. The coefficient of thermoelectromotive force of the films is 10-15 μV/deg C; the sign of the thermo-emf indicates n-type conduction.

6. The reduced glass layers satisfactorily obey Ohm's law over a wide range of field intensities (from 10^{-2} to 10^4 W/cm).

7. The relationship between the surface temperature of reduced glasses and the power dissipation is linear (up to 0.5 W/cm^2).

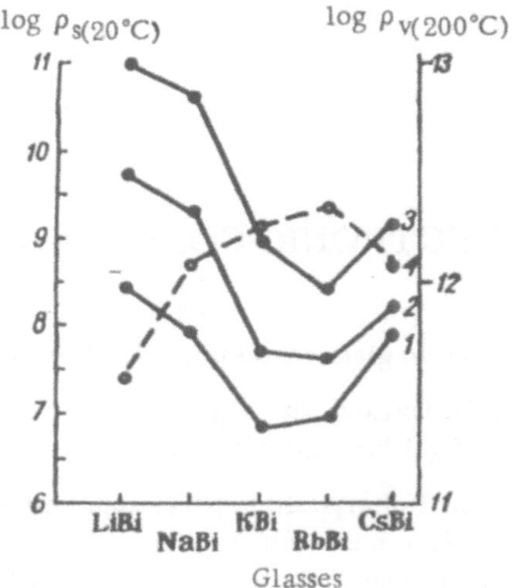

Fig. 1. Effects of various alkali-metal ions on the specific surface resistance of bismuth glasses reduced at: 1) 400°C; 2) 450°C; 3) 480°C. 4) Effects of alkali-metal ions on the specific volume resistance of bismuth glasses at 200°C.

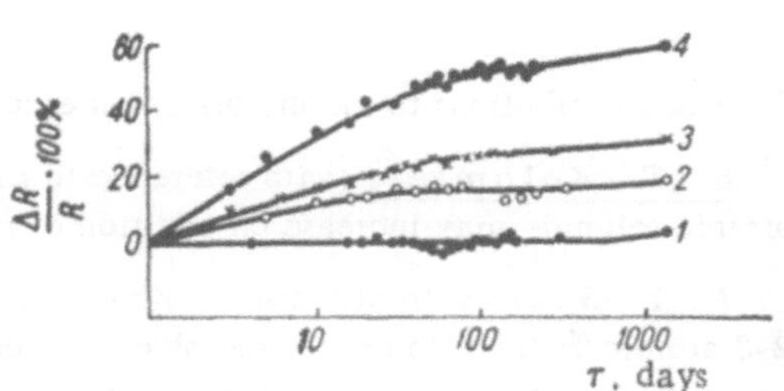

Fig. 2. Variations during storage of the surface resistance of lead glass reduced in hydrogen. Glass previously stabilized by heat treatment in air (1); glasses without previous stabilization: stored in a desiccator (2), with a protective film (3), and in air (4).

8. A series of experimental data (the dependence of resistance on temperature, the existence of thermo-electromotive force, and certain others) indicate electronic charge transfer in the reduced layers (layers 1-5 μ thick). We consider that during the first minutes of the reaction between hydrogen and the glass the reduced metal particles form uniformly distributed minute inclusions or "bridges." Subsequently these bridges partially aggregate into larger formations, consisting of pure metals or alloys having metallic conduction.* They have no effect on the film resistance, because they are either completely insulated from each other or are joined by bridges having higher resistance. It is the bridges which are the centers of conduction in the film.

Some investigators have regarded metal films applied by vacuum sputtering onto dielectric supports, such as glasses, as impurity semiconductors [5]. They considered that ions in the glass onto which the metal is sputtered may act as the impurities in such cases. By analogy, the bridges may be regarded as complex semiconductor systems with very low intrinsic and relatively high extrinsic conductivity. If it is assumed that bismuth, antimony, and arsenic ions act as donor impurities in the bridges, n-type conductivity should be observed; this is the case in reduced films.

It should be noted that this hypothesis of current transfer in reduced glasses is in good agreement with a number of experimental data omitted from this paper for reasons of space.

Literature Cited

1. R. Green and K. Blodgett, J. Am. Ceram. Soc., 31(4):89 (1948).
2. K. Blodgett, J. Am. Ceram. Soc., 34(1):14 (1951).
3. L. A. Grechanik, N. V. Solomin, E. A. Fainberg, and I. V. Shpakova, Scientific and Technical Collection of the Scientific Research Institute for Electrical Glass, No. 14 (1959) p. 51.
4. I. I. Kitaigorodskii, E. A. Fainberg, and L. A. Grechanik, Steklo i Keram. (12):8 (1962).
5. N. Mostovetch and B. Vodar, Semiconductor Materials [Russian translation], IL (1954) p. 338.

*The presence of these metallic aggregates has been conclusively proved by x-ray structural and differential thermal analysis.

DISCUSSION ON GLASSES WITH ELECTRONIC CONDUCTION

The contributions to the discussion on electronic conduction in glasses are summarized below.

B. T. Kolomiets (with reference to the paper by Z. U. Borisova and A. I. Bobrov). The conductivity of arsenic selenide may increase on addition of small amounts of gallium owing to crystallization of the glass.

A. I. Bobrov (reply to B. T. Kolomiets). A small amount of crystalline phase appeared on introduction of 2-3 atomic % Ga. No crystalline phase was detected in glass specimens with 1 atomic % Ga by x-ray diffraction investigation or by examination of sections under the metallographic microscope. Nevertheless, it was the first 1% of gallium which produced an appreciable increase of the conductivity modulus and brought it close to the theoretical value. It is therefore probable that the spatial network of covalent bonds is built up further by interaction of gallium with excess selenium in the glass. This should ensure better through transfer of the charge carriers along the covalent bonds.

V. F. Kokorina. Demonstration of the existence of crystalline inclusions in opaque chalcogenide glasses involves difficulties. X-ray structural analysis is unsatisfactory, as it does not always ensure detection of crystals, whereas the infrared transmission method gives definite indications of the presence of crystalline inclusions. These inclusions, which have a considerable effect on transmission, may possibly affect the conductivity less. Experience gained from numerous meltings of such glasses shows that transmission of at least 60% as shown by the infrared spectrometer is the only valid criterion of homogeneity and absence of crystalline inclusions in the glass.

A. M. Efimov. Contradictory interpretations of the part played by bond rupture are given in L. A. Baidakov's papers. For glasses in the system As—S this rupture, which leads to formation of readily ionizable radicals, is given as the cause of the decreased energy and increased conductivity. For glasses of the system As—Se—Ge, decrease of the conductivity modulus and fall of conductivity are attributed to the same cause. [Vestn. Leningr. Gos. Univ. (22):111 (1962)].

V. F. Kokorina. The suggestion made by R. L. Myuller and E. V. Shkol'nikov that nuclei form and crystallization subsequently proceeds within and on the surface of glasses of the arsenic—selenium—germanium system [Vestn. Leningr. Gos. Univ. (22):121 (1962)] gives rise to doubt. According to the extensive experience of the State Optical Institute, crystallization usually proceeds heterogeneously, starting at the glass surface and gradually penetrating inward. The views put forward by L. A. Baidakov on the structural chemistry of semiconducting glasses are not conclusive. Other structural arrangements are possible.

V. A. Khar'yuzov (with reference to the paper by L. A. Baidakov). Analysis of experimental data on the conductivity modulus, determined in the laboratories of B. T. Kolomiets, R. L. Myuller, and K. S. Evstrop'ev, and comparison of the results with the theoretical values, showed that for glasses of similar composition there are both agreements and differences between the experimental and theoretical values. Therefore agreement with the theoretical value of the conductivity modulus is no criterion of the validity of the experimental data. The discrepancies between the data of different authors are attributable either to experimental inaccuracies or to differences in the treatment of the glass.

O. V. Mazurin. Quantitative interpretation of data obtained by R. L. Myuller and E. V. Shkol'nikov on the effect of crystallization of $AsSe_{4.0}Ge_{5.0}$ and $AsSe_{2.33}Ge_{3.0}$ glasses on their conductivity is difficult. It is a very complicated matter to take into account the influence on electrical conductivity of the conducting crystalline inclusions distributed in the main medium of poorly conducting glass. These difficulties are increased by the difference in composition between the original glass and the crystals formed. For example, it is possible that a change of the glass composition (proportional to the amount of crystalline phase formed) predetermines a proportional change of energy. It would be advisable to investigate partially crystallized glasses in an alternating field in order to establish the relaxation maxima and to determine quantitatively the conducting crystalline phase and the dimensions of the blocked regions.

K. S. Evstrop'ev. Chalcogenide glasses, with well-defined structural groups, are more promising for investigation from the mechanical, physical, and chemical standpoints than oxide glasses. At the same time, somewhat more attention should be devoted to coordination changes in the structural formations of the glasses.

R. L. Myuller and L. A. Baidakov (reply to V. A. Khar'yuzov). There were no "experimental inaccuracies" in the cases discussed by V. A. Khar'yuzov. In fact, the reproducibility of the heat treatment before the conductivity determinations is poor. In particular, we see in the case of glass No. 15 ($AsSe_{3.73}Ge_{1.99}$) that the data on low-temperature conductivity marked "?" [Vestn. Leningr. Gos. Univ. (22):96, Table 1 (1962)] remained almost unchanged after repeated measurements at low temperatures. V. A. Khar'yuzov was able to obtain results in agreement with theory only after repeated annealing in high-temperature conductivity determinations. This case confirms once again that the theoretical calculations are meaningful. Our assertion that when such glasses are chilled bonds are broken in the network and are restored during subsequent annealing is also correct. This is supported by data on the influence of annealing in our work on crystallization.

R. L. Myuller and L. A. Baidakov (reply to A. M. Efimov). There is a difference between glasses of the As—S and As—Se—Ge systems. In the first the formation of a large number of biradicals in the glass volume, causing decrease of the energy ε_σ, is to be expected. In the second, bond periodicity is disturbed, with rupture of bonds and formation of radicals, at the interfaces of micropolymeric homogeneous ($AsSe_{3/2}$) and ($GeSe_{4/2}$) groups. The number of such breaks at the boundaries is relatively small and therefore they do not determine the nature of the principal carriers. At the same time, they break down through conduction and lower the conductivity modulus. These views can be verified by subsequent magnetic studies of the glasses in question.

E. V Shkol'nikov (reply to V. F. Kokorina). Comparable rates of surface and volume crystallization were attained as the result of heat treatment in which the effect of composition was taken into account, and with special determination of the optimum temperature range in the critical region.

E. V. Shkol'nikov (reply to O. V. Mazurin). The poorly conducting glassy component could no longer be decisive in specimens with more than 50% crystallization. Moreover, experimental data obtained immediately before the conference indicate that during crystallization of $AsSe_{1.5}$ glass specimens the conductivity gradually falls.

R. L. Myuller and E. V. Shkol'nikov (reply to V. F. Kokorina). The possibility of volume crystallization of $AsSe_{4.0}Ge_{5.0}$ and $AsSe_{2.33}Ge_{3.0}$ glasses is supported by the sharp difference between the course of the low-temperature variations of conductivity of these glasses (smooth decrease of the energy ε_σ) and the corresponding variations of surface-crystallizing glasses (abrupt change of ε_σ for glass to ε_σ for the crystals) [Vestn. Leningr. Gos. Univ. (22):120 (1962)].

R. L. Myuller (reply to V. F. Kokorina). The proposed structurochemical models of semiconducting glasses are, of course, tentative. However, such models are fruitful as a working hypothesis. Future accumulation of experimental material will be accompanied by revision and refinement of these structurochemical concepts. At the same time, the idea of a framework of covalent chemical bonds in glasses is already justified. It helps progressive scientific promotion of new experimental investigations and brings us nearer to a more rigorous theory of the structure of semiconducting glasses.

IV

ELECTRODE PROPERTIES OF GLASSES

GENERALIZED ION-EXCHANGE THEORY OF THE GLASS ELECTRODE

B. P. Nikol'skii and M. M. Shul'ts

In this paper we discuss certain aspects of the ion-exchange theory [1-5] which forms the basis of studies of the chemical nature of glasses by investigation of their electrode properties. In such investigations we start with the hypothesis that the electrode properties of glasses depend on the energy state of the mobile ions in their structure which are capable of exchanging with other ions in solution. As a result of such exchange, in particular, H^+ ions can penetrate into the glass from solution:

$$M^+_{gl} + H^+_{soln} \rightleftarrows H^+_{gl} + M^+_{soln} \tag{1}$$

In this case the difference between the energy states of the M^+ and H^+ ions in the glass and solution determines the electrode properties of the glass, i.e., its ability to respond by its electrode potential relative to the solution to changes in the activity of H^+ and M^+ ions in the solution. The theory was developed and refined as more experimental data became available.

For example, in the first variant of the theory (the "simple" theory) it was assumed that the bonding strengths of ions of a given kind were equal in all the ionizable groups of the glass, and that this bonding was independent of the position of equilibrium of reaction (1). From this an equation was derived for the dependence of the potential of a glass electrode on the activities of ions in solution:

$$\varphi = \varphi^0 + \vartheta \log(a_{H^+} + K a_{M^+}) \tag{2}$$

where $\vartheta = 2.3\ RT/F$ and K is the equilibrium constant of reaction (1) (exchange constant), which is equal to $\alpha_{H^+} N_{M^+} / \alpha_{M^+} N_{H^+}$, where α_i represents the activities of the ions in solution and N_i concentration in the glass.

Equation (2) describes the hydrogen ($\varphi = \varphi^0 + \vartheta \log a_{H^+}$) and metal ($\varphi = \varphi^0_{M^+} + \vartheta \log a_{M^+}$) functions of the glass and also the behavior of the glass in the region intermediate between the two functions.

The constant K in Eq. (2), which can be determined from experimental data on φ as a function of pH, is of considerable interest for characterization of the glass itself. The exchange constant K is a quantitative measure of the difference of the bonding strengths of H^+ and M^+ ions in the glass and in solution. For a given solvent, variations of K with the glass composition are determined by changes in the bonding strength of H^+ and M^+ ions in the glass. In many of our studies of the effect of glass composition on K it was shown how the ion bonding strengths vary with the chemical composition of the glasses, and the structural characteristics of the glasses were discussed on this basis [6-9]. However, it had already been observed in [1] and in other investigations that for a number of glasses the theoretical φ versus pH curves calculated from Eq. (2) diverge from the experimental data in the region intermediate between the hydrogen and metal functions.

To account for these discrepancies it was suggested that the ion bond strengths in the glass structure are nonuniform. This was taken into account quantitatively in [2], where the concept of "partial activities" of ions in different ionizable groups of the glass was used for deriving the equation of the "generalized" theory:

$$\varphi = \varphi^0 - \vartheta \log \sum_i \frac{\alpha_i \gamma_i}{a_{H^+} + \alpha_i K a_{M^+}} \tag{3}$$

where α_i is the ratio of the coefficients representing the bonding strengths of H^+ and M^+ ions in the i-th group in the glass, and γ_i is the mole fraction of that group.

It was shown in [2] that the theoretical φ versus pH curves for certain glasses, calculated from Eq. (3), are in satisfactory agreement with experimental data. A highly significant result of the generalized theory was the prediction of a possible stepwise course of the φ versus pH curves if the ionic bonds in the glass are strongly differentiated. This theoretical prediction was subsequently confirmed by experiment [7, 8]. However, for a number of reasons not discussed here, we subsequently introduced the concept of dissociation of the ionizable groups in the glass. Differences in ion bonding in the glass structure are reflected in different degrees of dissociation of these groups.

New equations were derived for the potential of the glass electrode from the general concepts of the theory and by application of the law of mass action to the dissociation of the ionizable groups. In particular, it was shown for glasses with one kind of weakly dissociating ionizable groups that the potential is given by the equation

$$\varphi = \varphi^0 + \frac{1}{2}\vartheta \log \left(a_{H^+} + Ka_{M^+}\right) + \frac{1}{2}\vartheta \log \left(a_{H^+} + \alpha Ka_{M^+}\right) \tag{4}$$

where α is the ratio of the dissociation constants of the ionizable groups in the H^+ and M^+ forms. Calculations showed that for binary glasses (M_2O-SiO_2) Eq. (4) gives better agreement with experimental φ versus pH curves than does Eq. (2) of the "simple" theory.

For glassy systems with H^+ and M^+ ions distributed between two kinds of ionizable groups (for example, alkali aluminosilicate glasses at low Al_2O_3/M_2O ratios) the relation between the potential and the ion activities is given by the expression

$$\varphi = \varphi^0 + \frac{1}{2}\vartheta \log \left(a_{H^+} + Ka_{M^+}\right) - \frac{1}{2}\vartheta \log \left(\frac{1}{a_{H^+} + \alpha_1 Ka_{M^+}} + \frac{\beta}{a_{H^+} + \alpha_2 Ka_{M^+}}\right) \tag{5}$$

where α_1 is the ratio of the dissociation constants of the first ionizable group in H^+ and M^+ forms; α_2 is the corresponding ratio for the second ionizable group; β is a coefficient which depends on the ratio of the concentrations of the ionizable groups and of the dissociation constants of their H^+ forms:

$$\beta = \frac{k_2^H N_2^0}{k_1^H N_1^0}$$

Equation (5) describes stepwise φ versus pH curves which are characteristic for glasses containing two glass-forming oxides, and gives satisfactory quantitative agreement with experimental data [5].

These equations can be used for calculating, from experimental data, the ion-exchange constant K which is determined by the nature of the glass and the solution, and the characteristic constants of the glass: the coefficients α_i, which depend on the dissociation constants of the ionizable groups, and the coefficient β, which takes into account the bonding strength of the H^+ ions in the ionizable groups and the total concentrations of these groups. Thus, as the result of development of the ion-exchange theory on the basis of the new concepts it is possible to go more deeply than before into the structural characteristics and chemical nature of glasses by studies of their electrode properties.

Literature Cited

1. B. P. Nikol'skii, Zh. Fiz. Khim., 10:495 (1937).
2. B. P. Nikol'skii, Zh. Fiz. Khim., 27:724 (1953).
3. B. P. Nikol'skii and M. M. Shul'ts, Zh. Fiz. Khim., 36:1327 (1962).
4. B. P. Nikol'skii and M. M. Shul'ts, Vestn. Leningr. Gos. Univ. (4):73 (1963).
5. B. P. Nikol'skii, M. M. Shul'ts, and A. A. Belyustin, Vestn. Leningr. Gos. Univ. (4):85 (1963).

6. B. P. Nikol'skii, N. P. Isakova, and M. M. Shul'ts, Dokl. Akad. Nauk SSSR, 142:1331 (1962).
7. M. M. Shul'ts, N. V. Peshekhonova, et al., Vestn. Leningr. Gos. Univ. (16):109 (1962); (4):120 (1963).
8. M. M. Shul'ts and A. A. Belyustin, Vestn. Leningr. Gos. Univ. (4):135 (1962); (16):116 (1962).
9. M. M. Shul'ts, Vestn. Leningr. Gos. Univ. (4):174 (1963).

ELECTRODE PROPERTIES AND CHEMICAL NATURE OF GLASS

M. M. Shul'ts, N. V. Peshekhonova, A. I. Parfenov,
A. A. Belyustin, and V. S. Bobrov

The electrode properties of glasses in numerous silicate systems have been systematically studied during recent years. The investigations included binary glasses (Li_2O-SiO_2 and Na_2O-SiO_2), three-component glasses containing, in addition to SiO_2 and Li_2O (or Na_2O), an oxide of one of the following elements: Cs, Be, Mg, Ca, Ba, Zn, Pb^{II}, B, Al, Ga, In, Sc, Y, La, Nd, Fe^{III}, Ge, Sn^{IV}, Ti, Zr, Ce^{IV}, Th, P^{V}, Sb^{III}, Bi^{III}, and certain four-component glasses. Variations of the potentials of electrodes made from these glasses with the pH of aqueous solutions (E versus pH curves) were studied at fixed Li^+ or Na^+ ion concentrations (0.1 or 3 N) under isothermal conditions (18, 60, or 95°C). These investigations are discussed in detail in a series of publications [1-8]. Here we give only a brief summary of the most important results.

The electrode properties of a glass depend in a definite manner on its chemical composition. The different effects of typical glass-forming and modifying oxides, the characteristics of intermediate oxides, and interaction of oxides of different types in the glass structure are clearly seen. The E versus pH curves for glasses of various systems are shown in Fig. 1. Curve 1 represents the electrode behavior of the original binary alkali silicate glass.

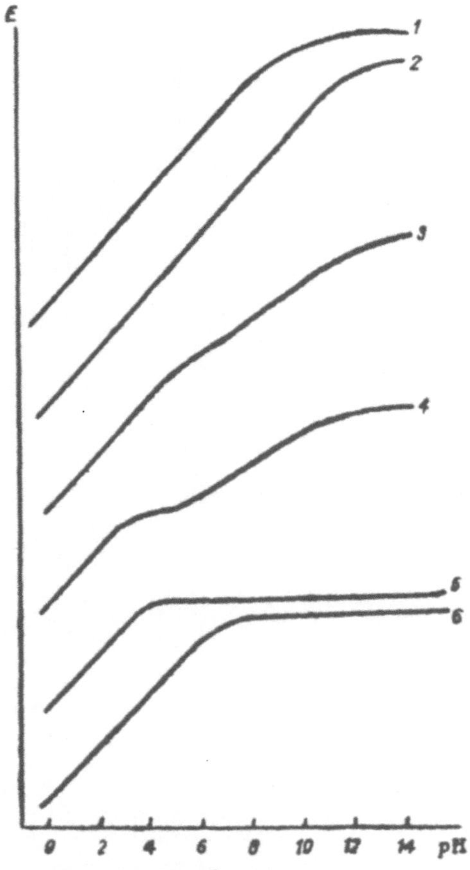

Fig. 1. Various types of E versus pH curves (see text).

Replacement of a small proportion of SiO_2 in alkali silicate glasses by oxides of the type R_2O_3, RO_2, and R_2O_5 results in a "step" or broad transitional region in the E versus pH curve between the hydrogen and metal functions (curves 3 and 4). Appearance of a "step" on the E versus pH curve when small amounts of oxides are added to the alkali silicate glass is due to formation of new structural units $[RO_{4/2}]^-$, $[RO_{6/2}]^{2-}$ in the glass network; these groups are negatively charged and are less strongly bonded to the hydrogen ions than in the original alkali silicate glass. However, the hydrogen ions bonded to the $SiO_{4/2}$ silicon–oxygen tetrahedrons still affect the electrode behavior of these glasses. This determines the different bonding strengths of H^+ ions (and M^+ ions) in the glass, dependent on the distribution of these ions between the strongly and weakly acidic silicate groups, and the effect appears on the E versus pH curve as a "step" [4].

With further introduction of R_2O_3 and RO_2 oxides the glass acquires a metallic electrode function over a wide range of pH (curve 5). The ion-exchange constant [9] increases sharply. The combination of these effects is defined as the electrode effect of the second glass-forming oxide.

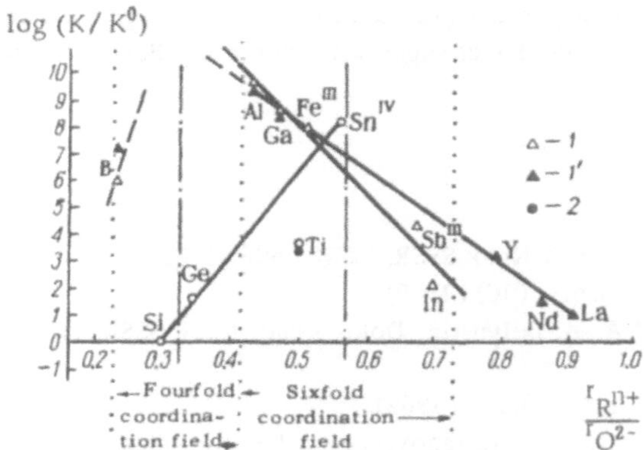

Fig. 2. Variation of log (K/K^0) with the ratio $r_{R^{n+}}/r_{O^{2-}}$. K – exchange constant for glass $M_2O-R_2O_3-SiO_2$; K^0 – exchange constant for the corresponding binary glass M_2O-SiO_2. 1) Glasses of the systems $Na_2O-R_2O_3-SiO_2$; 1') glasses of the systems $Li_2O-R_2O_3-SiO_2$ (presumed structural unit $[RO_{4/2}]^-$ for 1 and 1'); 2) glasses of the systems $Na_2O-R_2O-SiO_2$ (presumed structural unit $[RO_{6/2}]^{2-}$).

The extent of this effect depends primarily on the probability, based on the relative sizes of the R^{n+} and O^{2-} ions, of the formation of the ionizable units (groups). In particular, it is determined by the closeness of the $r_{R^{n+}}/r_{O^{2-}}$ ratio to the most favorable value for oxygen coordination of the element R^{n+}

It was found for R_2O_3 oxides that the closer the ratio $r_{R^{3+}}/r_{O^{2-}}$ to the most probable value for fourfold oxygen coordination of the R^{3+} ion the more prominent is the effect of strongly acidic $[RO_{4/2}]^-H^+$ groups in the electrode behavior of the glass. In the case of RO_2 oxides the effect of strongly acidic groups becomes more pronounced as the ratio $r_{R^{4+}}/r_{O^{2-}}$ approaches the value characteristic of sixfold oxygen coordination of the R^{4+} ion. This is confirmed by the plots of the exchange constant K versus the $r_{R^{n+}}/r_{O^{2-}}$ ratio, shown in Fig. 2. It is seen that the constant increases as the effect of the strongly acidic ionizable groups becomes more pronounced.

It was also found that the region of the hydrogen function extends somewhat and K decreases with increase of the $[R_2O_3]/[M_2O]$ or $[RO_2]/M_2O]$ ratio in the glass, for example with the oxides of Ga, In, Fe, Ti, etc. (curve 6, Fig. 1). This may be attributed to stronger bonding of H^+ ions in the glass structure due to partial removal of these elements from the glass network. Together with the effect, considered below, of ions having relatively low charge densities (Cs^+, Ba^{2+}, etc.), this effect is defined as the electrode effect of modifying ions. The appearance of such ions in the glass in places previously occupied by Li^+, Na^+, or H^+ ions may result in more complete capture of the remaining H^+ ions by the electron shells of oxygen atoms adjoining the modifier ions.

When modifying oxides M_2O and MO, which yield "foreign" modifying ions in the glass, are introduced into binary glasses (Li_2O-SiO_2 and Na_2O-SiO_2) the electrode behavior of the glasses does not show any sharp differentiation of the bonding strength of H^+ (or M^+) ions. However, it was found that with increasing contents of "foreign" M^+ and M^{2+} ions (and other cations from modifying oxides) in the glass structure the relative bonding strength of H^+ ions in the ionizable groups changes, increasing appreciably when the charge density of the modifying cation is relatively low, as in the case of Cs^+, Ba^{2+}, or Pb^{2+} (see curve 2, Fig. 1). However, this effect is most pronounced in presence of oxides which strengthen the glass structure and which are not typical glass-formers, such as La_2O_3, Nd_2O_3. This is especially prominent in the electrode behavior of glasses in four-component systems: $M_2O-Cs_2O(BaO)-La_2O_3-SiO_2$. In three-component glasses the effect of Cs^+, Ba^{2+}, and Pb^{2+} ions appears to be diminished somewhat owing to the considerable loosening of the glass structure.

The effect of Mg^{2+} and Be^{2+} ions on the electrode behavior of glasses is peculiar (curve 3, Fig. 1). On introduction of BeO or MgO into the glass the effect of formation of strongly acidic groups (possibly of the $RO_{4/2}]^{2-}$ type) can be detected. However, in this case there is no sharp differentiation of the bonding strengths of H^+ ions in the glass. The probability of formation of these strongly acidic groups is apparently low (especially with MgO). It is likely that in this case a wide range of ionizable groups, from strongly to weakly acidic, is formed in the glass structure.

The oxides Sc_2O_3, Y_2O_3 and rare-earth oxides have similar effects on the electrode behavior of glasses. Introduction of these oxides results in some weakening of the bonding of H^+ ions, which becomes more pronounced with decrease of the radius of the ion introduced. The formation of strongly acidic $[RO_{4/2}]^-H^+$ groups in the series of oxides Sc_2O_3, Y_2O_3, Nd_2O_3, La_2O_3 becomes increasingly more probable with decrease of the radius of the R^{3+} ion, so that K increases in the sequence from La_2O_3 to Sc_2O_3. In four-component systems containing two glass-formers and two modifiers, such as $Na_2O-BaO-Al_2O_3-SiO_2$ [7] or $Li_2O-BaO-ZrO_2-SiO_2$ [8, p. 143] we also observed that the influence of the second glass-former (Al_2O_3 or ZrO_2) is neutralized by the second modifier (BaO). The form of the E versus pH curves for these glasses gradually changes with increasing BaO concentration, from curves of type 5 to curves of type 2 (see Fig. 1).

Literature Cited

1. A. I. Parfenov, Vestn. Leningr. Gos. Univ. (4):98 (1959).
2. B. P. Nikol'skii, A. I. Parfenov, and M. M. Shul'ts, Dokl. Akad. Nauk SSSR, 127(3):599 (1959).
3. A. I. Parfenov and I. S. Ivanovskaya, Vestn. Leningr. Gos. Univ. (16) (1959).
4. B. P. Nikol'skii, M. M. Shul'ts, N. V. Peshekhonova, and A. A. Belyustin, Dokl. Akad. Nauk SSSR, 140:641 (1961).
5. M. M. Shul'ts and A. A. Belyustin, Vestn. Leningr. Gos. Univ. (4):135 (1962).
6. M. M. Shul'ts, N. V. Peshekhonova, L. M. Romanova, and A. A. Andrianov, Vestn. Leningr. Gos. Univ. (16):107 (1962).
7. M. M. Shul'ts and A. A. Belyustin, Vestn. Leningr. Gos. Univ. (16):116 (1962).
8. M. M. Shul'ts, N. V. Peshekhonova, A. I. Parfenov, et al., Vestn. Leningr. Gos. Univ., Ser. Fiz. i Khim. (4) (1963).
9. B. P. Nikol'skii and M. M. Shul'ts, this collection, p. 151.

DISCUSSION ON ELECTRODE PROPERTIES OF GLASSES

Some of the contributions to the discussion are summarized below.

K. K. Evstrop'ev. Recent work on the electrode properties of glasses reveals great potentialities for application of this method to studies of the structure of glass. However, these properties must be influenced by the state of the ions in the surface layers formed on the glass by interaction with solutions. It is necessary to elucidate the relationships between diffusion of the ions in the glass and their electric potential relative to the solution, and also the part played by leaching processes.

M. M. Shul'ts. In the general expression for the emf of a cell including a glass electrode two terms must be taken into account: the equilibrium interfacial potential at the glass–solution boundary and the diffusion potential within the glass membrane or surface film. It is at present difficult to specify the region of the glass to which the diffusion potential gradient should be referred. If ion diffusion in the surface film is great in comparison with diffusion within the glass and ion exchange "outpaces" hydration of the glass, the potential gradient is within the main bulk of the glass. In that case the glass–solution equilibrium potential is determined by the energy state of the ions in the glass structure. If the relation between the rates of the two processes is reversed, the potential is significantly influenced by the energy state of the ions in the modified surface film on the glass. Comparison of the behavior of glass electrodes in aqueous and nonaqueous media, in particular, is helpful for solving this problem. Such comparison shows that in some cases the solvent has a specific influence on the electrode potential, but the influence is relatively small. Therefore we can say that the energy state of the ions in the unchanged glass structure is the determining factor in the electrode properties of the glasses studied.

There is a close connection between such properties of glasses as conductivity, diffusion, and electrode properties. This is because the energy state of the mobile components of the glass structure – the alkali-metal ions – is very significant for all these properties. However, direct analogies should not be sought. The nature of the effects themselves must be examined in their interrelationships.

V. S. Molchanov. As M. M. Shul'ts points out, there is no perfect correlation between ion diffusion in glass, corrosion of glass, and the development of electrode properties. Glass corrosion is a very complex process. It consists of two essentially different stages: the first, a "prelude," leading to formation of hydroxides, and the second, where interaction occurs in the "aqueous" layer with the products of attack of the glass by hydroxides. The electrode properties of glasses are determined by processes of the first stage. Investigation of these properties may give useful information for understanding the structure of glass, but the results should not be extended beyond the possibilities of the method. For example, potash glasses, in which the mobility of the K^+ ion is low, are corroded rapidly; the process of glass corrosion is complex.

R. L. Myuller. If a glass has been corroded, the state of the intermediate layer is immaterial from the thermodynamic standpoint if equilibrium exists.

The extensive experimental data, not yet fully evaluated, indicate that the electrode properties of glasses are greatly dependent on their structure. The state of the ions in the glass structure determines its electrode properties.

K. S. Evstrop'ev. The papers on the electrode properties of glasses give results for silicate glasses only. Investigations of these properties in nonsilicate glasses are very important from the theoretical standpoint.

B. P. Nikol'skii. It is difficult to say at present which is the layer which determines thermodynamically the ionic equilibrium and therefore the potential difference between the glass and the solution. All the experimental results indicate an undoubted connection between the electrode properties of glasses and the state of ions in their structure. However, it is difficult to indicate the position of the layer which somehow retains the glass structure and is at the same time in equilibrium with the solution. The possibility is not excluded that the same intermediate layer determines a whole series of the physicochemical functions of the glass. However, it is a modified layer; its composition is not the same as within the glass. Otherwise only the corresponding metallic function (sodium for soda glass) would always appear. It is difficult to say at present whether this layer is also the film which helps to protect the glass against corrosion. Investigation of the electrode properties of nonsilicate glasses is indeed of both theoretical and practical interest. M. M. Shul'ts and V. S. Bobrov [Vestn. Leningr. Gos. Univ. (4) (1963)] showed that potassium aluminophosphate glasses and certain other nonsilicate glasses exhibit both hydrogen and metallic functions; i.e., they behave in many respects like silicate glasses.